Lecture Notes in Mathematics

A collection of informal reports and
Edited by A. Dold, Heidelberg and E

Series: Australian National Universit
Advisers: L. G. Kovács, B. H. Neumann, Hanna Neumann, M. F. Newman

220

W. A. Coppel
The Australian National University, Canberra/Australia

Disconjugacy

Springer-Verlag
Berlin · Heidelberg · New York 1971

AMS Subject Classifications (1970): 34 C 10, 34 A 30, 49 B 10

ISBN 3-540-05584-3 Springer-Verlag Berlin · Heidelberg · New York
ISBN 0-387-05584-3 Springer-Verlag New York · Heidelberg · Berlin

Offsetdruck: Julius Beltz, Hemsbach.

PREFACE

Disconjugacy has assumed growing significance in recent years and for this reason seemed a worthwhile topic for the study group in differential equations at the Australian National University. The group met almost once a week throughout 1970 and covered most aspects of the subject. I am grateful to the other regular participants, Dr A. Howe, Mr G.C. O'Brien and Mr A.N. Stokes, for their collaboration and support. The notes were prepared to provide a permanent record of the lectures. They follow a logical rather than a chronological order. For their preparation I accept full responsibility. In particular the decision to omit special results for third and fourth order equations was mine. I hope that by making the notes available in their present form they may prove useful to a wider audience. I thank Professor P. Hartman for his comments on the first draft of these notes. I also thank Mrs Barbara Geary for her careful typing of the manuscript, and Professor B.H. Neumann and Dr M.F. Newman for their assistance with the proof-reading.

W.A. Coppel
Department of Mathematics
Institute of Advanced Studies
Australian National University

CONTENTS

CHAPTER 0

INTRODUCTION

We give here the basic definitions and explain the division of the notes into chapters.

DEFINITION 1. *A linear differential equation of order* n

$$y^{(n)} + p_1(t)y^{(n-1)} + \dots + p_n(t)y = 0 \tag{1}$$

is said to be disconjugate on an interval I *if every nontrivial solution has less than* n *zeros on* I, *multiple zeros being counted according to their multiplicity.*

For example, the equation

$$y^{(n)} = 0$$

is disconjugate on any interval, since every nontrivial polynomial of degree less than n has less than n zeros.

It is shown in Chapter 3 that if the coefficients $p_k(t)$ $(k = 1, \dots, n)$ are continuous on an *open or half-open* interval I then (1) is disconjugate on I if, and only if, every nontrivial solution has less than n distinct zeros on I, without regard to their multiplicity.

What is so special about the zeros? According to the following proposition, the answer is nothing at all. Disconjugacy of a differential equation means the possibility of *interpolation* by the solutions of the equation.

PROPOSITION 1. *Suppose the coefficients* $p_k(t)$ $(k = 1, \dots, n)$ *are continuous on the interval* I. *Then the equation* (1) *is disconjugate on* I *if and only if for every* $m \le n$ *distinct points* $t_1, \dots, t_m$ *in* I, *for arbitrary positive integers* $r_1, \dots, r_m$ *with sum* $r_1 + \dots + r_m = n$ *and for* n *arbitrary real numbers* $\beta_1^{(1)}, \dots, \beta_1^{(r_1)}, \beta_2^{(1)}, \dots, \beta_m^{(r_m)}$ *there is a solution* $y(t)$ *such that*

$$y^{(\nu_j-1)}(t_j) = \beta_j^{(\nu_j)} \quad (\nu_j = 1, \dots, r_j;\ j = 1, \dots, m) .$$

Proof. Let $y_1(t), \dots, y_n(t)$ be linearly independent solutions of (1). Then all solutions of (1) have the form

$$y(t) = \alpha_1 y_1(t) + \dots + \alpha_n y_n(t) .$$

Write

$$A = \begin{bmatrix} y_1(t_1) & \cdots\, y_n(t_1) \\ \cdots\cdots & \\ y_1^{(r_1-1)}(t_1) & \cdots\, y_n^{(r_1-1)}(t_1) \\ y_1(t_2) & \cdots\, y_n(t_2) \\ \cdots\cdots & \\ y_1^{(r_m-1)}(t_m) & \cdots\, y_n^{(r_m-1)}(t_m) \end{bmatrix}, \quad \underset{\sim}{\beta} = \begin{bmatrix} \beta_1^{(1)} \\ \vdots \\ \beta_1^{(r_1)} \\ \beta_2^{(1)} \\ \vdots \\ \beta_m^{(r_m)} \end{bmatrix}, \quad \underset{\sim}{\alpha} = \begin{bmatrix} \alpha_1 \\ \alpha_2 \\ \vdots \\ \\ \vdots \\ \alpha_n \end{bmatrix}.$$

Then we want to choose $\underset{\sim}{\alpha}$ so that $A\underset{\sim}{\alpha} = \underset{\sim}{\beta}$. By linear algebra this is possible for *every* $\underset{\sim}{\beta}$ if and only if the corresponding homogeneous equation $A\underset{\sim}{\alpha} = 0$ has no nontrivial solution.

For systems of differential equations we use a different definition of disconjugacy.

DEFINITION 2. *Let* $A(t)$, $B(t)$, $C(t)$ *be* $n \times n$ *matrix functions, with* $A(t)$ *and* $C(t)$ *symmetric, and let* $B^*(t)$ *denote the transpose of* $B(t)$. *The linear Hamiltonian system of differential equations*

$$\begin{aligned} y' &= B(t)y + C(t)z \\ z' &= -A(t)y - B^*(t)z \end{aligned} \tag{2}$$

is said to be disconjugate on an interval I *if for every nontrivial solution* $y(t)$, $z(t)$ *the* n*-vector* $y(t)$ *vanishes at most once on* I.

The motive for considering systems of this particular form comes originally from mechanics and is further discussed at the beginning of Chapter 2. The following proposition can be proved in the same way as the previous one.

PROPOSITION 2. *Suppose the coefficients* $A(t)$, $B(t)$, $C(t)$ *are continuous on the interval* I. *Then the system* (2) *is disconjugate on* I *if and only if for every two distinct points* t_1, t_2 *in* I *and arbitrary* n*-vectors* η_1, η_2 *there is a solution* $y(t)$, $z(t)$ *such that*

$$y(t_1) = \eta_1\,, \quad y(t_2) = \eta_2\,.$$

It is shown in Chapter 2, §7, that a linear self-adjoint differential equation of order $2n$

$$(-1)^n\left[p_n(t)y^{(n)}\right]^{(n)} + (-1)^{n-1}\left[p_{n-1}(t)y^{(n-1)}\right]^{(n-1)} + \ldots + p_0(t)y = 0 \tag{3}$$

is equivalent to a system of the form (2). Accordingly the equation (3) is disconjugate in the sense of Definition 2 if every nontrivial solution has at most one zero of multiplicity $\geq n$. Thus if (3) is disconjugate in the sense of Definition 1 then it is disconjugate also in the sense of Definition 2, but for $n > 1$ the converse need not hold.

However, for the second order equation

$$[p(t)y']' + q(t)y = 0 , \tag{4}$$

which is equivalent to the system

$$y' = z/p(t)$$
$$z' = -q(t)y ,$$

both definitions coincide. The equation (4) is disconjugate if every nontrivial solution has at most one zero. Moreover, any second order equation

$$y'' + p_1(t)y' + p_2(t)y = 0$$

can be written in the self-adjoint form (4), with

$$p(t) = \exp\left[\int_{t_0}^{t} p_1(s)ds\right] , \quad q(t) = p_2(t)p(t) .$$

Because of their simplicity we begin by discussing second order equations in Chapter 1. Many of the results obtained there are generalized to linear Hamiltonian systems in Chapter 2 and, in a different direction, to equations of arbitrary order in Chapter 3. In Chapter 3 we use the first definition of disconjugacy and in Chapter 2 the second.

CHAPTER 1

SECOND ORDER EQUATIONS

1. General properties

We are going to study the equation

$$L[y] \equiv \big(p(t)y'\big)' + q(t)y = 0 \ , \tag{1}$$

where $p(t)$ and $q(t)$ are continuous functions on an interval I with $p(t) > 0$ and $q(t)$ real.

LEMMA 1. *Let $u(t)$, $v(t)$ be differentiable functions on the interval $[a, b]$ such that $u(a) = u(b) = 0$ and $v(t) \neq 0$ for $a \le t \le b$. Then there exists a constant α such that $u(t) + \alpha v(t)$ has a double zero in (a, b).*

Proof. Put $w = vu' - uv'$. Since $(u/v)' = w/v^2$ it follows from Rolle's theorem that $w(c) = 0$ for some c in (a, b). Hence there exist constants λ, μ, not both zero, such that

$$\lambda u(c) + \mu v(c) = 0$$

$$\lambda u'(c) + \mu v'(c) = 0 \ .$$

Since $v(c) \neq 0$ we must have $\lambda \neq 0$. Then $\alpha = \mu/\lambda$ has the required property.

Sturm's separation theorem is a simple consequence of this lemma.

PROPOSITION 1. *If $y_1(t)$, $y_2(t)$ are linearly independent solutions of the equation* (1) *then their zeros are interlaced, i.e. between two consecutive zeros of one there is exactly one zero of the other.*

Proof. Evidently $y_1(t)$ and $y_2(t)$ have no common zeros and no nontrivial linear combination of $y_1(t)$ and $y_2(t)$ has a double zero. Hence, by Lemma 1, $y_2(t)$ has at least one zero between two consecutive zeros of $y_1(t)$. Interchanging y_1 and y_2 we see that $y_2(t)$ has at most one zero between two consecutive zeros of $y_1(t)$.

COROLLARY. *Suppose the interval $I = [a, b)$ is half-open. If one nontrivial solution of* (1) *has infinitely many zeros on I then every solution has infinitely many zeros on I. If no nontrivial solution has infinitely many zeros on I then* (1) *is disconjugate on some subinterval $[c, b)$.*

The equation (1) is said to be *oscillatory* if every solution has infinitely many zeros on I.

THEOREM 1. *The equation* (1) *is disconjugate on the interval* I *if it has a solution without zeros on* I . *For a compact or an open interval* I *this condition is also necessary.*

Proof. The sufficiency of the condition follows at once from Proposition 1. We now prove its necessity and assume first that $I = [a, b]$ is compact. Let $y_1(t)$, $y_2(t)$ be solutions such that $y_1(a) = 0$, $y_1'(a) > 0$ and $y_2(b) = 0$, $y_2'(b) < 0$. Then, since (1) is disconjugate, $y_1(t) > 0$ for $a < t \le b$ and $y_2(t) > 0$ for $a \le t < b$. The solution $y(t) = y_1(t) + y_2(t)$ is positive for $a \le t \le b$.

Suppose next that $I = (a, b)$ is open. Let c be any point in I and choose a_ν, b_ν so that $a_\nu < c < b_\nu$, $a_\nu \downarrow a$ and $b_\nu \uparrow b$. Then there exists a solution $y_\nu(t)$ which is positive for $a_\nu \le t \le b_\nu$. By multiplying $y_\nu(t)$ by a suitable positive constant we can suppose that $y_\nu^2(c) + y_\nu'^2(c) = 1$. Letting $\nu \to \infty$ through a suitable subsequence we obtain $y_\nu(c) \to \alpha$, $y_\nu'(c) \to \beta$, where $\alpha^2 + \beta^2 = 1$. Then $y_\nu(t) \to y(t)$, $y_\nu'(t) \to y'(t)$ for all t in I , where $y(t)$ is the solution which satisfies the initial conditions $y(c) = \alpha$, $y'(c) = \beta$. Therefore $y(t) \ge 0$ for all t in I . If $y(\tau) = 0$ for some τ in I then also $y'(\tau) = 0$ and hence $y(t) \equiv 0$, which is a contradiction.

The existence of a solution without zeros is not necessary for disconjugacy on a *half-open* interval. For example, every nontrivial solution

$$y(t) = A\sin(t-\alpha) \quad (A > 0,\ 0 \le \alpha < 2\pi)$$

of the equation $y'' + y = 0$ vanishes exactly once on the interval $I = [0, \pi)$.

THEOREM 2. *The equation* (1) *is disconjugate on the half-open interval* $I = [a, b)$ *if it is disconjugate on its interior* (a, b) .

Proof. We need only show that if $y(t)$ is the solution satisfying the initial conditions $y(a) = 0$, $y'(a) = 1$ then $y(t) \ne 0$ for $a < t < b$. Suppose on the contrary that $y(c) = 0$, where $a < c < b$. Since $y'(c) \ne 0$, $y(t)$ assumes negative values in the neighbourhood of c . Therefore if $\varepsilon > 0$ is sufficiently small the solution $\tilde{y}(t)$ satisfying the initial conditions $\tilde{y}(a+\varepsilon) = 0$, $\tilde{y}'(a+\varepsilon) = 1$ has a zero near c . But this contradicts the hypothesis that (1) is disconjugate on (a, b) .

We show next that if the equation (1) has a solution without zeros the second order linear differential operator L can be factored into first order linear differential operators.

PROPOSITION 2. *Suppose the equation* (1) *has a solution* $u(t)$ *without zeros on the interval* I . *Then for any function* $y(t) \in C^2(I)$

$$L[y] \equiv L_1^*[pL_1[y]] , \tag{2}$$

where

$$L_1[y] = y' - ry , \quad L_1^*[y] = y' + ry$$

and $r(t) = u'(t)/u(t)$.

Proof. We have

$$pL_1[y] = py' - p(u'/u)y$$

and hence

$$\begin{aligned} L_1^*[pL_1[y]] &= (py')' - (pu')'y/u - pu'y'/u + p(u'/u)^2y + p(u'/u)y' - p(u'/u)^2y \\ &= L[y] - qy + qy \\ &= L[y] . \end{aligned}$$

This result can also be expressed in the equivalent form

$$L[y] \equiv u^{-1}D[pu^2D(y/u)] , \tag{3}$$

where $D = d/dt$. Proposition 2 may be compared with the algebraic result that any positive definite symmetric matrix A can be expressed in the form $A = T^*DT$, where D is a positive diagonal matrix and T is a triangular matrix with 1's in the main diagonal.

THEOREM 3. *The equation* (1) *has a solution without zeros on the interval* I *if and only if the Riccati equation*

$$R[w] \equiv w' + q(t) + w^2/p(t) = 0 \tag{4}$$

has a solution defined throughout I .

Proof. If $u(t)$ is a non-vanishing solution of (1) then $w(t) = p(t)u'(t)/u(t)$ is a solution of (4). Conversely if $w(t)$ is a solution of (4) any nontrivial solution of the first order linear equation

$$y' = [w(t)/p(t)]y$$

is also a non-vanishing solution of (1).

THEOREM 4. *If the Riccati inequality*

$$R[v] \equiv v' + q(t) + v^2/p(t) \leq 0 \tag{5}$$

has a solution $v(t)$ *defined throughout* I *then the equation* (1) *has a solution without zeros on* I .

Proof. Let c be an interior point of I and let $w(t)$ be the solution of

(4) such that $w(c) = v(c)$. By Theorem 3 it is sufficient to show that $w(t)$ is defined throughout I . By the theory of differential inequalities $v(t) \le w(t)$ for all $t > c$ for which $w(t)$ is defined. On the other hand, if

$$u(t) = v(c) - \int_c^t q(s)ds$$

then $R[u] = u^2/p \ge 0$ and hence $w(t) \le u(t)$ for all $t > c$ for which $w(t)$ is defined. It follows that $w(t)$ is defined and satisfies the inequalities $v(t) \le w(t) \le u(t)$ for all $t > c$ in I . Similarly $u(t) \le w(t) \le v(t)$ for all $t < c$ in I .

THEOREM 5. *Suppose the equation* (1) *is disconjugate on the half-open interval* $I = [a, b)$. *Then there exists a solution* $y_1(t)$ *such that for any linearly independent solution* $y_2(t)$

$$y_1(t)/y_2(t) \to 0 \quad as \quad t \to b \; . \tag{6}$$

Moreover

$$\int^b \frac{dt}{py_1^2} = \infty \, , \quad \int^b \frac{dt}{py_2^2} < \infty \, . \tag{7}$$

If $w_k = py_k'/y_k$ $(k = 1, 2)$ *then* $w_1(t) < w_2(t)$ *for all* t *near* b .

Proof. Let $u(t)$ and $v(t)$ be two linearly independent solutions of (1). Then

$$W = p(vu'-uv')$$

is a non-zero constant. Since $v(t)$ vanishes at most once and $W = pv^2(u/v)'$ it follows that u/v is monotonic near b . Therefore

$$\gamma = \lim_{t\to b} u(t)/v(t)$$

exists. If $\gamma = \pm\infty$ then by interchanging u and v we can make $\gamma = 0$. If γ is finite then by replacing u by $u - \gamma v$ we can make $\gamma = 0$. Thus by suitable choice of u and v we can always ensure that $\gamma = 0$. If we now take $y_1(t) = u(t)$ and $y_2(t) = \lambda v(t) + \mu u(t)$, where $\lambda \ne 0$, then

$$\frac{y_1(t)}{y_2(t)} = \frac{u(t)/v(t)}{\lambda+\mu u(t)/v(t)} \to 0 \quad \text{as} \quad t \to b \; .$$

From $(u/v)' = W/pv^2$ we get by integration

$$\gamma = u(c)/v(c) + W\int_c^b \frac{dt}{pv^2}, \tag{8}$$

whether γ is finite or not. For $u = y_1$, $v = y_2$ we have $\gamma = 0$ and hence the second relation (7) holds. For $u = y_2$, $v = y_1$ we have $\gamma = \pm\infty$ and hence the first relation (7) holds.

Choose c near b so that $y_1(t) \neq 0$, $y_2(t) \neq 0$ for $c \leq t < b$. Since w_k is unaltered if y_k is replaced by $-y_k$ we can suppose $y_1(t) > 0$, $y_2(t) > 0$. If we take $u = y_2$, $v = y_1$ then

$$w_1 - w_2 = -W/y_1y_2 .$$

But $W > 0$ by (8), since $\gamma = +\infty$. Hence $w_1(t) < w_2(t)$ for $c \leq t < b$.

Obviously the solution $y_1(t)$ is uniquely determined up to a constant factor. It is called the *principal solution* of (1). It is the solution which is 'smallest' near b. By the proof of Theorem 3, $w(t)$ is a minimal solution near b of the Riccati equation (4).

THEOREM 6. *Suppose we are given a sequence of equations*

$$L_\nu[y] \equiv (p_\nu(t)y')' + q_\nu(t)y = 0, \tag*{(1)$_\nu$}$$

each of which is disconjugate on the open or half-open interval I. *If as* $\nu \to \infty$

$$p_\nu(t) \to p(t) > 0, \quad q_\nu(t) \to q(t)$$

uniformly on compact subintervals of I *then the limit equation* (1) *is also disconjugate on* I.

Proof. For definiteness we will assume that I is open at its right endpoint. Suppose on the contrary that (1) has a nontrivial solution $y(t)$ with two zeros a, b $(a < b)$. Let $y_\nu(t)$ be the solution of $(1)_\nu$ such that $y_\nu(a) = 0$, $y'_\nu(a) = y'(a)$. Then $y_\nu(t) \to y(t)$ uniformly on compact subintervals of I. Since $y(t)$ changes sign at b it follows that for large ν, $y_\nu(t)$ vanishes near b, which is a contradiction.

The result need not hold if the interval I is compact. For example, the equation $y'' + \lambda^2 y = 0$ is disconjugate on the interval $[0, \pi]$ for $\lambda^2 < 1$ but not for $\lambda^2 = 1$.

2. Comparison theorems

Disconjugacy plays an important role in the calculus of variations, in the form of the Jacobi condition. This connection depends on the fact that if we set out to minimise the quadratic functional

$$Q[y] = \int_a^b (py'^2 - qy^2)\,dt \tag{9}$$

subject to the boundary conditions $y(a) = y(b) = 0$ then (1) is the corresponding Euler-Lagrange equation.

We will say that a function $y(t)$ is *admissible* on a compact interval $[a, b]$ if it is piecewise continuously differentiable and $y(a) = y(b) = 0$. We will say that the quadratic functional $Q[y]$ is *positive* if $Q[y] > 0$ for all nontrivial admissible functions $y(t)$.

In the calculus of variations the following identity is well-known.

LEMMA 2. *Let $u(t)$ be a twice differentiable function and $v(t)$ a differentiable function. If $y(t) = u(t)v(t)$ then*

$$py'^2 - qy^2 = pu^2v'^2 + (pu'vy)' - L[u]vy .$$

Proof. The right side is equal to

$$pu^2v'^2 + pu'(vy)' + (pu')'vy - (pu')'vy - quvy = pu^2v'^2 + pu'(uv^2)' - qy^2 .$$

This is the same as the left side, since

$$\begin{aligned} u^2v'^2 + u'(uv^2)' &= u^2v'^2 + 2uu'vv' + u'^2v^2 \\ &= (uv'+u'v)^2 \\ &= y'^2 . \end{aligned}$$

Taking $v(t) \equiv 1$ we obtain as a corollary the formula

$$Q[y] = [pyy']_a^b - \int_a^b yL[y]\,dt \tag{10}$$

for any twice continuously differentiable function $y(t)$. In particular

$$Q[y] = 0 \quad \text{if} \quad y(a) = y(b) = 0 \quad \text{and} \quad L[y] = 0 . \tag{11}$$

PROPOSITION 3. *Suppose there exists an admissible function $y(t)$ such that $Q[y] \le 0$. If $u(t)$ is a solution of (1) on the interval $[a, b]$ and if $y(t)$ is not a constant multiple of $u(t)$ then $u(c) = 0$ for some c in (a, b).*

Proof. Suppose on the contrary that $u(t) \ne 0$ for $a < t < b$ and set $v(t) = y(t)/u(t)$. By Lemma 2 if $a < a_1 < b_1 < b$ then

$$\int_{a_1}^{b_1} (py'^2-qy^2)dt = \int_{a_1}^{b_1} pu^2v'^2dt + [pu'vy]_{a_1}^{b_1} .$$

Now $pu'vy = pu'y^2/u \to 0$ as $t \to a$. This is obvious if $u(a) \neq 0$, since $y(a) = 0$. If $u(a) = 0$ then $u'(a) \neq 0$ and hence, by the theory of indeterminate forms,

$$\lim_{t\to a} pu'y^2/u = \lim_{t\to a} (2pu'yy'-quy^2)/u' = 0 .$$

Similarly $pu'vy \to 0$ as $t \to b$. Therefore, letting $a_1 \to a$, $b_1 \to b$ we obtain

$$Q[y] = \int_a^b pu^2v'^2dt .$$

Since the left side is nonpositive and the right side nonnegative their common value must be zero. Hence $v' = 0$ and $v = $ const., *i.e.*, $y(t)$ is a constant multiple of $u(t)$.

THEOREM 7. *The equation* (1) *is disconjugate on the compact interval* $I = [a, b]$ *if and only if the quadratic functional* $Q[y]$ *is positive.*

Proof. Suppose first that the functional is positive and let $u(t)$ be a solution of (1) such that $u(c) = u(d) = 0$, where $a \leq c < d \leq b$. Set $y(t) = u(t)$ for $c \leq t \leq d$, $= 0$ elsewhere. Then $y(t)$ is admissible and, by (11),

$$Q[y] = \int_c^d (pu'^2-qu^2)dt = 0 .$$

Therefore $y(t) \equiv 0$, and hence $u(t) \equiv 0$.

Suppose next that (1) is disconjugate and let $u(t)$ be the solution which satisfies the initial conditions $u(a) = 0$, $u'(a) = 1$. If $Q[y] \leq 0$ for some nontrivial admissible function $y(t)$ then by Proposition 3 either $u(c) = 0$ for some c in (a, b) or $u(t)$ is a constant multiple of $y(t)$ and hence $u(b) = 0$. In either case we have a contradiction.

Suppose now that we have two equations

$$L_1[y] = (p_1(t)y')' + q_1(t)y = 0 , \qquad (1)_1$$

$$L_2[y] = (p_2(t)y')' + q_2(t)y = 0 . \qquad (1)_2$$

From Theorem 7 we obtain at once the

COROLLARIES.

(i) If both equations $(1)_1$ *and* $(1)_2$ *are disconjugate on an interval* I *and if*

$$p(t) = \lambda_1 p_1(t) + \lambda_2 p_2(t) , \quad q(t) = \lambda_1 q_1(t) + \lambda_2 q_2(t) ,$$

where $\lambda_1 > 0$, $\lambda_2 > 0$, *then the equation* (1) *is also disconjugate on* I .

(ii) If the equation $(1)_2$ *is disconjugate on an interval* I *and if* $p_1(t) \geq p_2(t) > 0$ *and* $q_1(t) \leq q_2(t)$ *for all* t *in* I *then the equation* (1) *is also disconjugate on* I .

In these corollaries it is not necessary to assume that the interval I is compact, since disconjugacy on any interval is equivalent to disconjugacy on all compact subintervals. The second corollary is a form of the Sturm comparison theorem. A somewhat stronger form can be obtained by appealing directly to Proposition 3.

PROPOSITION 4. *Let* $y_1(t)$ *and* $y_2(t)$ *be solutions on the interval* $I = [a, b]$ *of the equations* $(1)_1$ *and* $(1)_2$ *respectively, where* $p_1(t) \geq p_2(t) > 0$ *and* $q_1(t) \leq q_2(t)$ *for all* t *in* I . *If* $y_1(a) = y_1(b) = 0$ *and if* $y_1(t)$ *is not a constant multiple of* $y_2(t)$ *then* $y_2(c) = 0$ *for some* c *in* (a, b) .

Proof. We have

$$\begin{aligned} Q_2[y_1] &= \int_a^b \left\{p_2 y_1'^2 - q_2 y_1^2\right\} dt \\ &\leq \int_a^b \left\{p_1 y_1'^2 - q_1 y_1^2\right\} dt \\ &= Q_1[y_1] \\ &= 0 , \end{aligned}$$

by (11). The result now follows at once from Proposition 3.

One can connect other variational problems with the differential equation (1) by adding to the functional $Q[y]$ a quadratic form in the values of y at the endpoints a, b and by varying the boundary conditions. We will consider only one simple case.

THEOREM 8. *Let* $I = [a, b]$ *be a compact interval. Then the following statements are equivalent:*

(i) *the equation* (1) *has no nontrivial solution* $u(t)$ *such that* $u'(b) = u(c) = 0$ *for some* c *in* $[a, b)$,

(ii) *the inequality* $yL[y] \leq 0$ *has a solution* $u(t) \neq 0$ *on* I *with* $p(b)u'(b)u(b) \geq 0$,

(iii) *the quadratic functional* $Q[y]$ *is positive definite on the set of*

piecewise continuously differentiable functions $y(t)$ *such that* $y(a) = 0$.

Proof. *(i)* ⇒ *(ii)*. Let $u(t)$ be the solution of (1) such that $u(b) = 1$, $u'(b) = 0$. Then $u(t) > 0$ for $a \le t \le b$ and *(ii)* holds.

(ii) ⇒ *(iii)*. For any piecewise continuously differentiable function $y(t)$ such that $y(a) = 0$, set $v(t) = y(t)/u(t)$. Then by Lemma 2

$$\begin{aligned} Q[y] &= \int_a^b pu^2v'^2dt + [pu'vy]_a^b - \int_a^b v^2uL[u]dt \\ &\ge \int_a^b pu^2v'^2dt \\ &\ge 0 , \end{aligned}$$

with equality only if $v'(t) = 0$, *i.e.*, $y(t)/u(t) = \text{const.}$ Since $y(a) = 0$ it follows that $y(t) = 0$.

(iii) ⇒ *(i)*. Let $u(t)$ be a solution of (1) such that $u'(b) = u(c) = 0$ and put $y(t) = 0$ for $a \le t \le c$, $= u(t)$ for $c \le t \le b$. Then, by (10),

$$Q[y] = \int_c^b (pu'^2 - qu^2)dt = 0 .$$

Therefore $y(t) \equiv 0$ and $u(t) \equiv 0$.

Proposition 4 can also be proved without reference to the calculus of variations. If y_1, y_2 are nontrivial solutions of $(1)_1$ and $(1)_2$ respectively and $y_2 \ne 0$ then

$$\begin{aligned} [(y_1/y_2)(p_1y_1'y_2 - p_2y_1y_2')]' &= y_1(p_1y_1')' - (y_1^2/y_2)(p_2y_2')' + (p_1-p_2)y_1'^2 + p_2(y_1'-y_1y_2'/y_2)^2 \\ &= (q_2-q_1)y_1^2 + (p_1-p_2)y_1'^2 + p_2(y_1'-y_1y_2'/y_2)^2 . \end{aligned} \qquad (12)$$

Let a, b be zeros of y_1 and suppose $y_2 \ne 0$ on the open interval (a, b) . If $y_2(a) \ne 0$ then evidently

$$\lim_{t\to a} (y_1/y_2)(p_1y_1'y_2 - p_2y_1y_2') = 0 .$$

If $y_2(a) = 0$ then $y_2'(a) \ne 0$ and again

$$\begin{aligned} \lim_{t\to a} (y_1/y_2)[p_1y_1'y_2 - p_2y_1y_2'] &= \lim_{t\to a} [y_1'(p_1y_1'y_2 - p_2y_1y_2') + y_1(p_1y_1'y_2 - p_2y_1y_2')']/y_2' \\ &= 0 . \end{aligned}$$

Similarly

$$\lim_{t\to b} (y_1/y_2)[p_1y_1'y_2 - p_2y_1y_2'] = 0 .$$

Hence, integrating (12) over the interval $[a, b]$ we get

$$\int_a^b \left[(q_2-q_1)y_1^2 + (p_1-p_2)y_1'^2 + p_2(y_1'-y_1y_2'/y_2)^2\right]dt = 0 .$$

If $p_1 \geq p_2 > 0$ and $q_1 \leq q_2$ it follows that $(y_1/y_2)' \equiv 0$ and y_1 is a constant multiple of y_2 on $[a, b]$.

A simple proof of Sturm's comparison theorem in its most general form is provided by the method of polar coordinates. If we put

$$y = r\sin\theta , \quad py' = r\cos\theta ,$$

the equation (1) is replaced by the pair of first order equations

$$\theta' = p^{-1}\cos^2\theta + q\sin^2\theta , \tag{13}$$

$$r'/r = (p^{-1}-q)\sin\theta\cos\theta . \tag{14}$$

Hence all nontrivial solutions of the equation (1) on $I = [a, b]$ are given by

$$y = Cr(t)\sin\theta(t) , \quad py' = Cr(t)\cos\theta(t) ,$$

where $\theta(t)$ is a solution of the equation (13) with $0 \leq \theta(a) < \pi$, C is a nonzero constant and

$$r(t) = \exp \tfrac{1}{2} \int_a^t [p^{-1}(s)-q(s)]\sin 2\theta(s)ds .$$

Since $r(t) > 0$ the zeros of y are given by $\theta(t) = j\pi$, where j is an arbitrary integer.

Suppose now that we have two equations, $(1)_1$ and $(1)_2$, where $p_1(t) \geq p_2(t) > 0$ and $q_1(t) \leq q_2(t)$. The corresponding angular coordinate equations are

$$\theta' = f_k(t, \theta) = p_k^{-1}\cos^2\theta + q_k\sin^2\theta \quad (k = 1, 2) .$$

Thus $f_1(t, \theta) \leq f_2(t, \theta)$. If $\theta_1(a) \leq \theta_2(a)$ then by the theory of differential inequalities $\theta_1(b) \leq \theta_2(b)$ with equality only if $\theta_1(t) = \theta_2(t)$ for $a \leq t \leq b$. The conditions for equality can be made more explicit. We must have

$$f_2(t, \theta) - f_1(t, \theta) = \left(p_2^{-1}-p_1^{-1}\right)\cos^2\theta + (q_2-q_1)\sin^2\theta = 0 .$$

Hence on an interval on which $q_2 \neq q_1$ we must have $\sin\theta = 0$; therefore $\theta' = p_k^{-1} = 0$, which is impossible. Similarly on an interval on which $p_2 \neq p_1$ we must have $\cos\theta = 0$, therefore $\theta' = q_k = 0$. Thus $\theta_1(b) < \theta_2(b)$ if

$\theta_1(a) < \theta_2(a)$

OR $\theta_1(a) = \theta_2(a)$ and for at least one point of the interval I

$$\text{either } p_1 > p_2 \text{ and } q_1^2 + q_2^2 > 0 \text{ , or } q_2 > q_1 \text{ .} \tag{15}$$

THEOREM 9. *Let* $y_1(t), y_2(t)$ *be solutions of the equations* $(1)_1, (1)_2$ *respectively, where* $p_1(t) \geq p_2(t) > 0$ *and* $q_1(t) \leq q_2(t)$. *If either* $y_1(a) = 0$ *or* $y_1(a) \neq 0$, $y_2(a) \neq 0$ *and* $p_1(a)y_1'(a)/y_1(a) \geq p_2(a)y_2'(a)/y_2(a)$ *then* $y_2(t)$ *has at least as many zeros as* $y_1(t)$ *in the interval* $a < t \leq b$. *If* t_n', t_n'' *are the n-th zeros of* $y_1(t), y_2(t)$ *then* $t_n'' \leq t_n'$. *Moreover* $t_n'' < t_n'$ *if* (15) *holds for at least one point of the interval* $[a, t_n']$.

Proof. Let $\theta_1(t), \theta_2(t)$ be the corresponding angular coordinates. Initially $\sin\theta_1(a) = 0$ or $\cot\theta_1(a) \geq \cot\theta_2(a)$, *i.e.*, $0 \leq \theta_1(a) \leq \theta_2(a) < \pi$. Therefore, by what has been said above, $\theta_1(t) \leq \theta_2(t)$ with strict inequality if (15) holds for at least one point of the interval $[a, t]$. If $\theta_k(t) = j\pi$ then $\theta_k' = p_k^{-1} > 0$. That is, θ_k is increasing at this point and therefore passes through it only once. Since y_k vanishes if and only if $\theta_k = j\pi$ for some integer j , the result follows.

As another example of the application of the polar coordinate method we prove

PROPOSITION 5. *Suppose the equation* (1) *is disconjugate on the half-open interval* $I = [a, b)$. *Further let* $q(t) \geq 0$ *on* I , $q(t) \not\equiv 0$ *near* b *and*

$$\int^b dt/p(t) = \infty \text{ .}$$

If $y(t)$ *is a solution of* (1) *such that* $y(a) = 0$, $y'(a) > 0$ *then* $y'(t) > 0$ *for* $a < t < b$.

Proof. If we put

$$y = r\sin\theta \text{ , } py' = r\cos\theta \quad (r > 0) \text{ ,}$$

the angular coordinate $\theta(t)$ of $y(t)$ is the solution of the equation

$$\theta' = p^{-1}(t)\cos^2\theta + q(t)\sin^2\theta$$

such that $\theta(a) = 0$. Since $q(t) \geq 0$ it follows that $\theta(t)$ is a non-decreasing function of t . Therefore, since (1) is disconjugate, we must have $0 < \theta(t) < \pi$ for $a < t < b$. If $\theta(c) > \pi/2$ then for $t > c$

$$\theta(t) - \theta(c) > \int_c^t p^{-1}\cos^2\theta ds \geq \cos^2\theta(c) \int_c^t p^{-1} ds .$$

Since $\int^b p^{-1}(t)dt = \infty$ this gives a contradiction for t near b. Therefore $\theta(t) \leq \pi/2$ for $a < t < b$. If $\theta(c) = \pi/2$ then $\theta(t) = \pi/2$ for $t \geq c$. Hence $\theta'(t) = q(t) = 0$, contrary to the hypothesis that $q(t)$ does not vanish for all t near b. Therefore $0 < \theta(t) < \pi/2$ and $y'(t) > 0$ for $a < t < b$.

Thus, under the conditions of the proposition, there is no nontrivial solution $y(t)$ such that $y(c) = y'(d) = 0$ $(a \leq c < d < b)$. Therefore, by Theorem 8,

$$\int_a^d (py'^2 - qy^2)dt > 0$$

for any piecewise continuously differentiable function $y(t)$ with $y(a) = 0$. If we take

$$\begin{aligned} y(t) &= \int_a^t p^{-1}(s)ds \Big/ \int_a^c p^{-1}(s)ds \quad \text{for } a \leq t \leq c \\ &= 1 \quad \text{for } t \geq c , \end{aligned}$$

it follows that

$$\int_c^d q dt \leq \int_a^d qy^2 dt < \left[\int_a^c p^{-1} dt\right]^{-1} .$$

Hence, under the conditions of Proposition 5, $\int^b q(t)dt$ exists and

$$\int_a^t p^{-1}(s)ds \int_t^b q(s)ds \leq 1 \quad \text{for } a \leq t < b .$$

We establish next an integral comparison test for disconjugacy. In the proof we will use the following

LEMMA 3. *Let $p(t) > 0$ be a continuous function on $I = [a, b)$ such that $\int^b p^{-1}(t)dt = \infty$ and let $w(t)$ be a continuous function such that*

$$v(t) = w(t) + \int_a^t p^{-1}(s)w^2(s)ds$$

is bounded above. Then

$$\int_a^b p^{-1}(t)w^2(t)dt < \infty .$$

If $v(t)$ has a finite limit as $t \to b$ then $w(t) \to 0$ as $t \to b$.

Proof. Suppose on the contrary that

$$\int_a^t p^{-1}(s)w^2(s)ds \to \infty \quad \text{as} \quad t \to b .$$

If $v(t) \leq K$ for $t \geq a$ then, setting

$$W(t) = \int_a^t p^{-1}(s)w^2(s)ds - K ,$$

we have $W(t) > 0$ for $t \geq c > a$. Since $W(t) \leq -w(t)$ it follows that

$$W^2 \leq w^2 = pW' \quad \text{for} \quad t \geq c .$$

Hence

$$\int_c^t p^{-1}(s)ds \leq \int_c^t W^{-2}W'ds \leq W^{-1}(c) ,$$

which gives a contradiction for t sufficiently close to b.

It follows that if v has a finite limit as $t \to b$ then w has also. Since $\int_a^b p^{-1}w^2dt < \infty$ and $\int_a^b p^{-1}dt = \infty$ we must have $w \to 0$.

THEOREM 10. *Given two equations,* $(1)_1$ *and* $(1)_2$*, on the interval* $I = (a, b)$ *such that the integrals*

$$Q_k(t) = \int_t^b q_k(s)ds \quad (k = 1, 2)$$

converge and $\int^b p_2^{-1}(t)dt = \infty$. *Suppose* $p_1(t) \geq p_2(t) > 0$ *and* $|Q_1(t)| \leq Q_2(t)$ *for all* t *in* I. *Then if the equation* $(1)_2$ *is disconjugate on* I, *so also is the equation* $(1)_1$.

Proof. Since $(1)_2$ is disconjugate on the open interval I the Riccati equation

$$w' + q_2(t) + w^2/p_2(t) = 0$$

has a solution $w(t)$ defined throughout I. Integrating, we get

$$w(c) + \int_t^c p_2^{-1}(s)w^2(s)ds = w(t) - \int_t^c q(s)ds . \tag{16}$$

The right side has a finite limit as $c \to b$, and hence the left side has also. Therefore, by Lemma 3,

$$\int_t^b p_2^{-1}(s)w^2(s)ds < \infty$$

and $w(c) \to 0$ as $c \to b$. Thus, letting $c \to b$ in (16) we obtain

$$w(t) = \int_t^b p_2^{-1}(s)w^2(s)ds + Q_2(t) .$$

Put

$$v(t) = \int_t^b p_2^{-1}(s)w^2(s)ds + Q_1(t) .$$

Then $|v(t)| \le w(t)$ and

$$v'(t) = - q_1(t) - w^2(t)/p_2(t)$$
$$\le - q_1(t) - v^2(t)/p_1(t) .$$

Therefore, by Theorem 4, the equation $(1)_1$ is disconjugate on I.

3. Tests for oscillation

In this section we depart from our usual practice and give oscillation precedence over disconjugacy.

THEOREM 11. *The equation* (1) *is oscillatory on the interval* $I = [a, b)$ *if* $\int^b p^{-1}dt = \infty$ *and there exists a continuously differentiable function* $u(t) > 0$ *such that*

$$\int_a^b (qu^2-pu'^2)dt = + \infty .$$

Proof. Suppose on the contrary that (1) is disconjugate on some interval $[c, b)$. For a suitable non-principal solution $y(t)$ we have $y(t) > 0$ for $t \ge c$ and

$$\int_c^b dt/py^2 < \infty .$$

Then $w(t) = p(t)y'(t)/y(t)$ is a solution of the Riccati equation (4). If we multiply (4) by u^2 and integrate from c to t we get

$$\int_c^t u^2w'dt + \int_c^t u^2qdt + \int_c^t u^2w/pdt = 0 .$$

Therefore

$$[u^2w]_c^t = -\int_c^t (p^{-1/2}uw-p^{1/2}u')^2dt - \int_c^t (qu^2-pu'^2)dt$$

$$\leq -\int_c^t (qu^2-pu'^2)dt .$$

From the definition of $w(t)$ it follows that

$$y'(t)/y(t) < 0 \quad \text{for} \quad t \geq T \text{ , say.}$$

Thus $y(t)$ is a decreasing function for $t \geq T$ and

$$\int_T^b p^{-1}dt < y^2(T) \int_T^b dt/py^2 < \infty ,$$

which is a contradiction.

For example, if $p(t) \equiv 1$ and if we take $u(t) = t^{\alpha/2}$ or $t^{1/2}(\log t)^{-(1+\beta)/2}$ we obtain the

COROLLARY. *The equation*

$$y'' + q(t)y = 0$$

is oscillatory if

$$\int^{\infty} t^{\alpha}q(t)dt = +\infty \quad \textit{for some} \quad \alpha < 1$$

or if

$$\int^{\infty} t \log^{-1-\beta}t \; q(t)dt = +\infty \quad \textit{for some} \quad \beta > 0 .$$

PROPOSITION 6. *The equation* (1) *is oscillatory on the interval* $I = [a, b)$ *if* $\int^b p^{-1}dt = \infty$ *and* $\int^b qdt = \infty$.

This is another corollary of Theorem 11 $(u \equiv 1)$. It can also be proved in the following way.

If the equation (1) is nonoscillatory on I then it is disconjugate in the neighbourhood of b and the Riccati equation (4) has a solution $w(t)$ defined on a subinterval $[c, b)$. By integration we get

$$w(t) - w(c) + \int_c^t q(s)ds + \int_c^t p^{-1}(s)w^2(s)ds = 0 .$$

It follows from Lemma 3 that $\int_c^b p^{-1}w^2dt < \infty$. Therefore $w(t) \to -\infty$ as $t \to b$ and hence $\int_c^b p^{-1}dt < \infty$, which is a contradiction.

THEOREM 12. *The equation* (1) *is oscillatory on the interval* $I = [a, b)$ *if and only if there exists a twice continuously differentiable function* $u(t) > 0$ *such that*

$$\int^b dt/pu^2 = +\infty , \quad \int^b uL[u]dt = +\infty . \tag{17}$$

Proof. Let y_1, y_2 be two linearly independent solutions of (1) such that

$$p(y_1y_2'-y_2y_1') \equiv 1 .$$

Then $u = \left(y_1^2+y_2^2\right)^{1/2}$ is a solution of the nonlinear equation

$$(pu')' + qu - (pu^3)^{-1} ,$$

and all solutions of the original equation (1) are given by

$$y(t) = Au(t)\sin\left\{\int_a^t ds/pu^2 + \alpha\right\} ,$$

where A and α are arbitrary constants. The equation (1) is nonoscillatory on I if and only if

$$\int_a^b dt/pu^2 < \infty$$

and is disconjugate on I if and only if

$$\int_a^b dt/pu^2 < \pi .$$

Thus if (1) is oscillatory on I then

$$\int_a^c dt/pu^2 = \int_a^c uL[u]dt \to \infty \quad \text{as} \quad c \to b .$$

Conversely, suppose there exists $u(t) > 0$ for which (17) holds. The change of variables $y = u(t)z$ transforms (1) into the equation

$$(pu^2z')' + uL[u]z = 0 ,$$

which is oscillatory by Proposition 6. It follows that (1) is also oscillatory.

4. The case $p(t) \equiv 1$

For the equation

$$y'' + q(t)y = 0 \tag{18}$$

it is possible to obtain a number of more specialised results. For example, it is immediately verified that the autonomous equation

$$y'' + \lambda y = 0$$

is disconjugate on $(-\infty, \infty)$ if $\lambda \leq 0$ and oscillatory on $[0, \infty)$ if $\lambda > 0$. We can deduce from this that the Euler equation

$$y'' + \lambda t^{-2} y = 0$$

is disconjugate on $(0, \infty)$ if $\lambda \leq 1/4$ and oscillatory on $[1, \infty)$ if $\lambda > 1/4$. In fact the change of variables

$$T = e^t \ , \quad Y = e^{t/2} y \tag{19}$$

transforms the equation (18) into an equation

$$d^2Y/dT^2 + Q(T)y = 0$$

of the same form, where

$$Q(T) = [\tfrac{1}{4} + q(\log T)]/T^2 \ .$$

The same transformation applied to the Euler equation shows that the equation

$$y'' + \left(\frac{1}{4t^2} + \frac{\lambda}{t^2 \log^2 t}\right) y = 0$$

is disconjugate on $(1, \infty)$ if $\lambda \leq 1/4$ and oscillatory on $[e, \infty)$ if $\lambda > 1/4$. And so on. Combined with the Sturm comparison theorem, or the integral comparison test, this yields a sequence of more and more refined criteria for disconjugacy and oscillation.

Again, the equation

$$y'' + \lambda y = 0$$

is disconjugate on the interval $I = [a, b]$ if $\lambda < \pi^2/(b-a)^2$. It follows from the Sturm comparison theorem that the equation (18) is disconjugate on $[a, b]$ if

$$\sup_{a \leq t \leq b} q(t) < \pi^2/(b-a)^2 \ .$$

In particular the equation (18) is disconjugate on a compact interval if the coefficient $q(t)$ is sufficiently small in the L^∞-norm. Our next result shows that the equation (18) is disconjugate on a compact interval if the coefficient

$q(t)$ is sufficiently small in the L^1-norm.

THEOREM 13. *Let* $q(t) \geq 0$ *be continuous on the compact interval* $I = [a, b]$. *If*

$$\int_a^b q(t)dt \leq 4/(b-a) \qquad (20)$$

then the equation (18) *is disconjugate on* I .

Proof. We will show that if (18) has a nontrivial solution $y(t)$ with two zeros on I then (20) cannot hold. Since $q(t) \geq 0$ we can assume that the zeros are at the endpoints a, b . The inhomogeneous equation

$$y'' + f(t) = 0$$

can be integrated directly. The solution which vanishes at a and b is given by

$$\int_a^b G(t, s)f(s)ds ,$$

where

$$G(t, s) = \begin{cases} (b-t)(s-a)/(b-a) & \text{for } a \leq s \leq t \\ (t-a)(b-s)/(b-a) & \text{for } t \leq s \leq b . \end{cases}$$

It follows that $y(t)$ satisfies the integral equation

$$y(t) = \int_a^b G(t, s)q(s)y(s)ds .$$

Choose t so that

$$|y(t)| = \sup_{a \leq s \leq b} |y(s)| > 0 .$$

Then we obtain

$$1 \leq \int_a^b G(t, s)q(s)ds .$$

Since

$$0 \leq G(t, s) < (b-s)(s-a)/(b-a) \quad \text{for } s \neq t$$

it follows that

$$b - a < \int_a^b (b-s)(s-a)q(s)ds .$$

Since

$$(b-s)(s-a) \leq (b-a)^2/4 \quad \text{for} \quad a \leq s \leq b$$

we obtain finally

$$4/(b-a) < \int_a^b q(s)ds \ .$$

The constant 4 in Theorem 13 is best possible. For suppose $0 < \delta < \frac{1}{2}$ and let $y(t)$ be a twice continuously differentiable function on the interval $I = [0, 1]$ such that

$$y(t) = t \qquad \text{for} \quad 0 \leq t \leq \tfrac{1}{2}-\delta \ ,$$

$$y(t) = 1 - t \quad \text{for} \quad \tfrac{1}{2}+\delta \leq t \leq 1 \ ,$$

$$y(t) > 0,\ y''(t) < 0 \quad \text{for} \quad \tfrac{1}{2}-\delta < t < \tfrac{1}{2}+\delta \ .$$

If we set $q(t) = -y''(t)/y(t)$ for $0 < t < 1$, $q(0) = q(1) = 0$, then $q(t)$ is nonnegative and continuous on I. Moreover

$$\begin{aligned}\int_0^1 q(t)dt &= -\int_{\frac{1}{2}-\delta}^{\frac{1}{2}+\delta} [y''(t)/y(t)]dt \\ &\leq \left(\tfrac{1}{2}-\delta\right)^{-1}\left[y'\left(\tfrac{1}{2}-\delta\right) - y'\left(\tfrac{1}{2}+\delta\right)\right] \\ &= 4(1-2\delta)^{-1} \ .\end{aligned}$$

The equation (18) is not disconjugate on I, but $\int_0^1 q(t)dt$ can be made arbitrarily close to 4 by taking δ sufficiently small.

The equation (18) is disconjugate on the interval $[a, \infty)$ if and only if the corresponding Riccati equation

$$w' + q(t) + w^2 = 0 \tag{21}$$

has a solution $w(t)$ defined throughout the open interval (a, ∞). This places a restriction on the coefficient $q(t)$. As we will see, it means that as $t \to \infty$,

$$\int_a^t q(s)ds$$

either almost converges to a finite limit or almost diverges to $-\infty$. In order to make precise what we mean by 'almost' here we introduce the following definition.

Let $f(t)$ be a continuous function defined for $t \geq a$. We say that $f(t)$ *converges to a finite limit* λ *in* L^2 *as* $t \to \infty$ if we can write

$$f(t) = \lambda + g(t) + h(t)$$

where $g(t)$, $h(t)$ are continuous functions and $g(t) \to 0$ as $t \to \infty$, $h(t) \in L^2(a,\infty)$.

Evidently the limit λ is uniquely determined.

LEMMA 4. *Let* $f(t)$ *be a continuous function defined for* $t \geq a$ *such that* $f(t) \to \lambda$ *in* L^2 *as* $t \to \infty$. *If for each* $\varepsilon > 0$ *there exist* $\delta > 0$ *and* $T \geq a$ *such that*

$$f(t'') - f(t') \geq -\varepsilon \quad \textit{for} \quad T \leq t' \leq t'' \leq t'+\delta \,, \tag{22}$$

then $f(t) \to \lambda$ *in the ordinary sense as* $t \to \infty$.

Proof. It is sufficient to show that if $f(t) \in L^2(a, \infty)$ and satisfies (22) then $f(t) \to 0$ as $t \to \infty$. Suppose on the contrary that there exists a sequence $t_\nu \to \infty$ such that $f(t_\nu) \geq \eta > 0$. Taking $\varepsilon = \frac{1}{2}\eta$ we have

$$f(t) \geq f(t_\nu) - \tfrac{1}{2}\eta \geq \tfrac{1}{2}\eta \quad \text{for} \quad T \leq t_\nu \leq t \leq t_\nu+\delta \,,$$

which is impossible if $f(t) \in L^2(a, \infty)$. Similarly if there exists a sequence $t_\nu \to \infty$ such that $f(t_\nu) < -\eta < 0$ then

$$f(t) \leq f(t_\nu) + \tfrac{1}{2}\eta \leq -\tfrac{1}{2}\eta \quad \text{for} \quad T \leq t_\nu-\delta \leq t \leq t_\nu \,,$$

which again gives a contradiction.

Since f can be replaced by $-f$ the condition (22) can be replaced by

$$f(t'') - f(t') \leq \varepsilon \quad \text{for} \quad T \leq t' \leq t'' \leq t'+\delta \,.$$

Both conditions are satisfied if $f(t)$ is uniformly continuous, and one of them is satisfied if $f(t)$ is differentiable and its derivative is semi-bounded.

PROPOSITION 7. *Suppose the Riccati equation* (21) *has a solution* $w(t)$ *on the interval* $[a, \infty)$. *Then* $w(t) \in L^2(a, \infty)$ *if and only if* $\int_a^t q(s)ds$ *has a finite limit* λ *in* L^2 *as* $t \to \infty$. *Moreover*

$$\lambda = w(a) - \int_a^\infty w^2(s)ds \,.$$

If $w(t) \notin L^2(a, \infty)$ *then*

$$\int_a^t q(s)ds \Big/ \int_a^t w^2(s)ds \to -1 \quad \textit{in} \quad L^2 \quad \textit{as} \quad t \to \infty \,.$$

In particular, for any real number μ *the set of all* t *such that* $\int_a^t q(s)ds > \mu$ *has finite measure.*

Proof. Put

$$W(t) = \int_a^t w^2(s)ds \ , \quad Q(t) = \int_a^t q(s)ds \ .$$

From (21) we obtain, by integration,

$$w(t) - w(a) + Q(t) + W(t) = 0 \ ,$$

that is

$$Q(t)/W(t) = -1 + w(a)/W(t) - w(t)/W(t) \ .$$

Evidently $w(a)/W(t) \to w(a)/W(\infty)$ as $t \to \infty$. Also $w(t)/W(t) \in L^2(b, \infty)$ for any $b > a$, since

$$(w/W)^2 = W'/W^2 = -(1/W)' \ .$$

Therefore the ratio $Q(t)/W(t)$ converges to a finite limit in L^2 as $t \to \infty$, namely $-1 + w(a)/W(\infty)$.

If $W(\infty) < \infty$ it follows that $Q(t)$ converges to the finite limit $\lambda = w(a) - W(\infty)$ in L^2 as $t \to \infty$. Conversely, if $Q(t)$ converges to a finite limit λ in L^2 as $t \to \infty$ then $Q(t)/W(t) \to \lambda/W(\infty)$ in L^2. Therefore $\lambda/W(\infty) = -1 + w(a)/W(\infty)$, which implies $W(\infty) < \infty$.

Suppose $W(\infty) = \infty$. If $Q(t) > \mu$ and t is large then

$$Q(t)/W(t) > \mu/W(t) > -\tfrac{1}{4}$$

and $w(a)/W(t) < \frac{1}{4}$. Therefore

$$w(t)/W(t) < -1 + \tfrac{1}{4} + \tfrac{1}{4} = -\tfrac{1}{2} \ .$$

Since $(w/W) \in L^2$ this can hold only on a set of finite measure.

It follows from Proposition 7 that either all or none of the solutions of (21) which are defined for large t have integrable square. It follows also that the equation (18) is oscillatory on the interval $[a, \infty)$ if $\int_a^t q(s)ds \to +\infty$ as $t \to \infty$, in agreement with Proposition 6.

LEMMA 5. *Let* $f(t)$ *be a continuous function for* $t \geq a$ *such that, for any real number* μ, *the set of all* t *such that* $f(t) > \mu$ *has finite measure. If* $f(t+\theta) - f(t)$ *is semi-bounded for* $t \geq a$, $0 \leq \theta \leq 1$, then

$$f(t) \to -\infty \quad as \quad t \to \infty \ .$$

Proof. Suppose on the contrary that there exists a sequence $t_\nu \to \infty$ such that $f(t_\nu) \geq \alpha > -\infty$. If $f(t+\theta) - f(t) \geq -\beta$ for $t \geq a$, $0 \leq \theta \leq 1$, then

$$f(t_\nu+\theta) \geq \alpha - \beta \quad \text{for} \quad 0 \leq \theta \leq 1 \quad \text{and all} \quad \nu \ .$$

If $f(t+\theta) - f(t) \le \beta$ for $t \ge a$, $0 \le \theta \le 1$ then

$$f(t_\nu - \theta) \ge \alpha - \beta \text{ for } 0 \le \theta \le 1 \text{ and all } \nu .$$

In either case we have a contradiction.

The condition of the lemma is certainly satisfied if $f(t)$ is a differentiable function and its derivative is semi-bounded. Thus from Proposition 7 and Lemmas 4 and 5 we obtain

PROPOSITION 8. *If the equation* (18) *is disconjugate on the interval* $I = [a, \infty)$ *and* $q(t)$ *is semi-bounded, then* $\int_a^t q(s)ds$ *either converges to a finite limit or diverges to* $-\infty$ *as* $t \to \infty$.

This result will be used in the proof of

THEOREM 14. *If* $q(t) \not\equiv 0$ *is almost periodic with mean value zero then the equation* (18) *is oscillatory.*

The proof will be conducted in a series of stages:

(i) If (18) is not disconjugate on $(-\infty, \infty)$ then it is oscillatory at $\pm\infty$.

For suppose there exists a solution $y(t)$ such that $y(a) = y(b) = 0$ $(a < b)$. There exist arbitrarily large positive and negative τ_n such that

$$|q(t) - q(t+\tau_n)| < 1/n \text{ for } -\infty < t < \infty .$$

Let $\tilde{y}_n(t)$ be the solution of the translated equation $y'' + q(t+\tau_n)y = 0$ which satisfies the same initial conditions at a as $y(t)$. Then, given any δ $(0 < \delta < b-a)$, $\tilde{y}_n(t)$ vanishes between $b - \delta$ and $b + \delta$ for all large n. But $\tilde{y}_n(t) = y_n(t+\tau_n)$, where $y_n(t)$ is a solution of the original equation (18). Thus $y_n(t)$ vanishes at $a + \tau_n$ and near $b + \tau_n$. Therefore, by Proposition 1, every solution of (18) vanishes between $a + \tau_n$ and $b + \tau_n + \delta$. Since this holds for all n, the equation (18) is oscillatory at $\pm\infty$.

(ii) If (18) is disconjugate on $(-\infty, \infty)$ then so is the equation $y'' + q^*(t)y = 0$ for any almost periodic function $q^*(t)$ in the closed hull of $q(t)$.

Since $q^*(t)$ is the uniform limit of a sequence of translates $q(t+\tau_n)$ this follows at once from Theorem 6.

(iii) If (18) is disconjugate on $(0, \infty)$ then $Q(t) = \int_0^t q(s)ds \to -\infty$ as $t \to \infty$.

Since $q(t)$ is bounded, Proposition 8 shows that $Q(t) \to \lambda$ as $t \to \infty$, where $-\infty \le \lambda < \infty$. If λ is finite then $Q(t)$ is bounded on $(0, \infty)$ and therefore almost periodic. Thus the almost periodic function $Q(t) - \lambda$ converges to zero as $t \to \infty$, which is possible only if it is identically zero. Therefore $q(t) = 0$, which has been excluded. Hence $\lambda = -\infty$.

(iv) If an almost periodic function $q(t)$ has mean value zero and $\int_0^t q(s)ds$ is unbounded there exists a function $q^*(t)$ in the closed hull of $q(t)$ such that $\int_0^t q^*(s)ds \ge 0$ for $-\infty < t < \infty$.

This is proved in Favard [1]. We can now complete the proof of the theorem. If (18) is not oscillatory then, by (i) and (ii), it and all equations in its closed hull are disconjugate on $(-\infty, \infty)$. Therefore, by (iii), $\int_0^t q^*(s)ds \to -\infty$ as $t \to \infty$, for any $q^*(t)$ in the closed hull of $q(t)$. However this contradicts (iv).

In Proposition 7 we have studied the conditions under which a solution of the Riccati equation (21) belongs to $L^2(a, \infty)$. We study next the conditions under which a solution converges to zero as $t \to \infty$.

PROPOSITION 9. *Suppose the Riccati equation* (21) *has a solution* $w(t)$ *on the interval* $[a, \infty)$. *Then* $w(t) \to 0$ *as* $t \to \infty$ *if and only if*

$$r(t) \equiv \sup_{t \le t_1 \le t_2 \le t_1 + 1} \left| \int_{t_1}^{t_2} q(s)ds \right| \to 0 \quad as \quad t \to \infty . \tag{23}$$

Proof. The necessity of (23) is trivial, since

$$\left| \int_{t_1}^{t_2} q(s)ds \right| \le |w(t_2) - w(t_1)| + \int_{t_1}^{t_2} w^2(s)ds .$$

Suppose conversely that (23) holds. Then

$$Q(t) = \int_a^t q(s)ds = o(t) \quad \text{for} \quad t \to \infty .$$

If $|w(t)| \ge \gamma > 0$ for all large t then

$$W(t) = \int_a^t w^2(s)ds \ge \tfrac{1}{2}\gamma^2 t \quad \text{for all large} \quad t ,$$

and hence $Q(t)/W(t) \to 0$ as $t \to \infty$. On the other hand, by Proposition 7, $Q(t)/W(t) \to -1$ in L^2 as $t \to \infty$. From this contradiction we conclude that

$\underline{\lim}|w(t)| = 0$.

If $\overline{\lim}|w(t)| \neq \underline{\lim}|w(t)|$ choose α, β so that

$$\underline{\lim}|w(t)| < \alpha < \beta < \overline{\lim}|w(t)| .$$

There exist intervals $[t_1, t_2]$, with t_1, t_2 arbitrarily large, such that

$$\alpha \leq |w(t)| \leq \beta \quad \text{for} \quad t_1 \leq t \leq t_2$$

and

$$w(t_2) - w(t_1) = \beta - \alpha .$$

Thus

$$-\int_{t_1}^{t_2} q(s)ds = \int_{t_1}^{t_2} w^2(s)ds + w(t_2) - w(t_1)$$

$$\geq \alpha^2(t_2-t_1) + \beta - \alpha .$$

But

$$\left|\int_{t_1}^{t_2} q(s)ds\right| \leq r(t_1)(t_2-t_1) + r(t_1) ,$$

which gives a contradiction if t_1 is so large that $r(t_1) < \alpha^2, \beta - \alpha$.

Although we have not used the fact in the proof it is worth noting that if the Riccati equation (21) has a solution on $[a, \infty)$ and (23) holds then, by Proposition 7 and Lemmas 4 and 5, $\int_a^t q(s)ds$ either converges to a finite limit or diverges to $-\infty$ as $t \to \infty$.

The Riccati equation (21) can also be reformulated as an integral equation.

LEMMA 6. *Let* $q(t)$ *be a continuous function for* $t \geq a$ *such that, for some constant* λ ,

$$Q(t) = \lambda - \int_a^t q(s)ds \to 0 \quad in \ L^2 \quad as \quad t \to \infty .$$

Then the Riccati equation (21) *has a solution* $w(t)$ *defined on* $[a, \infty)$ *if and only if the integral equation*

$$w(t) = Q(t) + \int_t^\infty w^2(s)ds \tag{24}$$

has a continuous solution $w(t)$ *on* $[a, \infty)$.

Proof. If $w(t)$ is a continuous solution of (24) then it is differentiable and is a solution of (21). Conversely, if (21) has a solution $w(t)$ then $w(t) \in L^2(a, \infty)$, by Proposition 7. Integrating (21), we get

$$w(t) = w(a) - \int_a^\infty w^2(s)ds - \int_a^t q(s)ds + \int_t^\infty w^2(s)ds ,$$

which coincides with (24) because λ is uniquely determined.

As an application we prove

THEOREM 15. *Suppose*

$$Q(t) = \lambda - \int_a^t q(s)ds \to 0 \quad \textit{in } L^2 \quad \textit{as} \quad t \to \infty \tag{25}$$

and $Q(t) \geq 0$ *for* $t \geq a$. *If*

$$\int_t^\infty Q^2(s)ds \leq \tfrac{1}{4}Q(t) \quad \textit{for} \quad t > a , \tag{26}$$

then the equation (18) *is disconjugate on* $[a, \infty)$.

If $Q(t) \not\equiv 0$ *for large* t *and*

$$\int_t^\infty Q^2(s)ds \geq (\tfrac{1}{4}+\varepsilon)Q(t) \quad \textit{for} \quad t > a , \tag{27}$$

where $\varepsilon > 0$, *then the equation* (18) *is oscillatory on* $[a, \infty)$.

Proof. Put

$$v(t) = Q(t) + 4\int_t^\infty Q^2(s)ds .$$

If (26) holds then $v^2 \leq 4Q^2$. Moreover v is differentiable and

$$v' = -q - 4Q^2 \leq -q - v^2 .$$

Therefore, by Theorem 4, the equation (18) is disconjugate on $[a, \infty)$.

Suppose conversely that (18) is not oscillatory on $[a, \infty)$. Then the Riccati equation (21) has a solution $w(t) \in L^2(b, \infty)$ for some $b \geq a$. Hence $w(t)$ is a solution of (24), which implies $w(t) \geq Q(t)$, for $t \geq b$. Since $Q(t) \not\equiv 0$ for large t there exists a greatest $\mu \geq 1$ such that $v(t) \geq \mu Q(t)$ for $t \geq b$. Then, if (27) holds,

$$\begin{aligned} w(t) &\geq Q(t) + \mu^2 \int_t^\infty Q^2(s)ds \\ &\geq Q(t) + (\tfrac{1}{4}+\varepsilon)\mu^2 Q(t) . \end{aligned}$$

Therefore $\mu \geq 1 + \left(\frac{1}{4}+\varepsilon\right)\mu^2$, which is impossible for $\varepsilon > 0$.

It will now be shown that if the equation (18) is disconjugate and (25) holds then $Q(t)$ can't tend to zero from above 'too slowly'.

THEOREM 16. *Suppose the equation* (18) *is nonoscillatory on the interval* $[a, \infty)$ *and* (25) *holds. Then*

$$\int^{\infty} \exp\left[-4\int_a^t Q_+(s)ds\right]dt = \infty , \tag{28}$$

where $Q_+(t) = \max[Q(t), 0]$.

Proof. If $y(t)$ is any nontrivial solution of (18) there exists $b \geq a$ such that $y(t) \neq 0$ for $t \geq b$. Then for $t \geq b$, $w(t) = y'(t)/y(t)$ is a solution of the Riccati equation (21) and

$$w(t) = Q(t) + \int_t^{\infty} w^2(s)ds .$$

Put

$$u(t) = \int_t^{\infty} w^2(s)ds .$$

Then $u(t)$ is differentiable and

$$\begin{aligned} u' &= -(u+Q)^2 \\ &\leq -2u|Q| - 2uQ \\ &= -4uQ_+ . \end{aligned}$$

Therefore

$$0 \leq u(t) \leq u(b)\exp\left[-4\int_b^t Q_+(s)ds\right] .$$

Thus if (28) does not hold then

$$\int_b^{\infty} u(t)dt < \infty .$$

Since $u(t)$ is a positive decreasing function this implies that $tu(t) \to 0$ as $t \to \infty$. Hence if we integrate by parts we get

$$\int_b^{\infty} tw^2(t)dt < \infty .$$

Choose $t_0 \geq b, e$ so large that

$$\int_{t_0}^{\infty} tw^2(t)dt \le \tfrac{1}{2} .$$

Then by Schwarz's inequality

$$\left[\int_{t_0}^{t} w(s)ds\right]^2 \le \int_{t_0}^{t} sw^2(s)ds \int_{t_0}^{t} ds/s$$

$$\le \tfrac{1}{2}\int_{t_0}^{t} s^{-1} \log s \, ds$$

$$\le \tfrac{1}{4} \log^2 t .$$

Since $w = y'/y$ it follows that

$$y^2(t)/y^2(t_0) \le t .$$

Hence

$$\int^{\infty} dt/y^2(t) = \infty .$$

But if we take $y(t)$ to be any non-principal solution of (18) this gives a contradiction.

Using Theorem 16 we next obtain another integral equation equivalent to the Riccati equation (21).

THEOREM 17. *Suppose* (25) *holds and put*

$$K(t, s) = \exp\left[2 \int_t^s Q(\tau)d\tau\right] .$$

If the equation (18) *is disconjugate on* (a, ∞) *then the integral equation*

$$u(t) = \int_t^{\infty} K(t, s)\left[Q^2(s)+u^2(s)\right]ds \tag{29}$$

has a continuous solution $u(t)$ *on* (a, ∞) .

Conversely, if the integral inequality

$$v(t) \ge \int_t^{\infty} K(t, s)\left[Q^2(s)+v^2(s)\right]ds \tag{30}$$

has a continuous solution $v(t)$ *on* (a, ∞) *then the equation* (18) *is disconjugate on* (a, ∞) .

Proof. Suppose first that (18) is disconjugate on (a, ∞) . Then by Lemma 6 the integral equation (24) has a continuous solution $w(t)$ on (a, ∞) . If we set

$$u(t) = \int_t^\infty w^2(s)ds ,$$

then $w(t) = u(t) + Q(t)$ and hence

$$u' = - (u+Q)^2 .$$

Writing this as a linear equation

$$u' + 2Qu = - (Q^2+u^2)$$

and solving in the usual way we obtain for $a < t < t_1$

$$u(t) = K(t, t_1)u(t_1) + \int_t^{t_1} K(t, s)[Q^2(s)+u^2(s)]ds .$$

Since the first term on the right and the integrand of the second term are non-negative it follows that

$$\int_t^\infty K(t, s)[Q^2(s)+u^2(s)]ds < \infty$$

and

$$k(t) = \lim_{t_1\to\infty} K(t, t_1)u(t_1)$$

exists and is finite. We wish to show that $k(t) = 0$ for $t > a$. Assume on the contrary that $k(t) > 0$ for some t. Choose T so large that

$$K(t, s)u(s) \geq \tfrac{1}{2}k(t) > 0 \quad \text{for} \quad s \geq T .$$

Since

$$\int_t^\infty K(t, s)u^2(s)ds < \infty ,$$

it follows that

$$\int_t^\infty K(s, t)ds < \infty ,$$

that is

$$\int_t^\infty \exp\left[- 2 \int_t^s Q(\tau)d\tau\right]ds < \infty ,$$

which contradicts Theorem 16.

Suppose next that the integral inequality (30) has a continuous solution $v(t)$ for $t > a$. Define a sequence $u_n(t)$ of continuous functions by setting $u_0(t) \equiv 0$

and

$$u_{n+1}(t) = \int_t^{\infty} K(t, s)\left[Q^2(s)+u_n^2(s)\right]ds \quad (n = 0, 1, \ldots) .$$

It is easily shown by induction that

$$0 \le u_n(t) \le u_{n+1}(t) \le v(t) .$$

Hence $u(t) = \lim u_n(t)$ exists and, by the theorem of monotone convergence, $u(t)$ is a solution of the integral equation (29). Therefore $u(t)$ is continuous and even differentiable and

$$u' = -\left(Q^2+u^2\right) - 2Qu = -(Q+u)^2 .$$

Therefore $w(t) = u(t) + Q(t)$ is a solution of the Riccati equation (21), and (18) is disconjugate on (a, ∞).

We obtain finally an analogue of Theorem 15.

THEOREM 18. *Suppose* (25) *holds and put*

$$\overline{Q}(t) = \int_t^{\infty} K(t, s)Q^2(s)ds ,$$

where

$$K(t, s) = \exp\left[2 \int_t^s Q(\tau)d\tau\right] .$$

If

$$\int_t^{\infty} K(t, s)\overline{Q}^2(s)ds \le \tfrac{1}{4}\overline{Q}(t) < \infty \quad \text{for} \quad t > a \tag{31}$$

then the equation (18) *is disconjugate on* (a, ∞).

If $Q(t) \not\equiv 0$ *for large* t *and*

$$\int_t^{\infty} K(t, s)\overline{Q}^2(s)ds \ge \left(\tfrac{1}{4}+\varepsilon\right)\overline{Q}(t) \quad \text{for} \quad t > a , \tag{32}$$

where $\varepsilon > 0$, *then the equation* (18) *is oscillatory on* (a, ∞).

Proof. If (31) holds then $v(t) = 2\overline{Q}(t)$ satisfies (30), and so the first part of the theorem follows at once from the second part of Theorem 17.

Suppose next that (32) holds. If (18) is not oscillatory, the integral equation (29) has a continuous solution $u(t)$ on (b, ∞) for some $b \ge a$. Then $u(t) \ge \overline{Q}(t)$ for $t \ge b$. Arguing as in the proof of Theorem 15 we obtain a contradiction.

Theorem 18 may be used to show that the equation

$$y'' + (\alpha t^{-1}\sin\beta t)y = 0 \quad (\alpha \neq 0,\ \beta \neq 0)$$

is oscillatory on $[1, \infty)$ if $(\alpha/\beta)^2 > \frac{1}{2}$ and nonoscillatory if $(\alpha/\beta)^2 < \frac{1}{2}$.

CHAPTER 2

LINEAR HAMILTONIAN SYSTEMS

1. General properties

A linear Hamiltonian system has the form

$$Jx' = H(t)x \ , \tag{1}$$

where J is a constant, invertible skew-symmetric matrix and $H(t)$ is a symmetric matrix function which is continuous on an interval I . Since J is skew-symmetric and invertible it must be of even order, $2n$ say. Moreover, by a linear change of variables we can, and will, assume that it has the form

$$J = \begin{pmatrix} 0 & -I_n \\ I_n & 0 \end{pmatrix} .$$

Let

$$x = \begin{pmatrix} y \\ z \end{pmatrix} , \quad H = \begin{pmatrix} A & B^* \\ B & C \end{pmatrix}$$

be the corresponding partitions of the vector x and the symmetric matrix H . Then A and C are symmetric $n \times n$ matrices and (1) is equivalent to the system

$$\begin{aligned} y' &= B(t)y + C(t)z \\ z' &= -A(t)y - B^*(t)z \ . \end{aligned} \tag{2}$$

It is convenient to consider at the same time the matrix equations

$$\begin{aligned} Y' &= B(t)Y + C(t)Z \\ Z' &= -A(t)Y - B^*(t)Z \ . \end{aligned} \tag{3}$$

LEMMA 1. *For any two solutions* Y_1, Z_1 *and* Y_2, Z_2 *of* (3) *the 'Wronskian'*

$$Y_1^*(t)Z_2(t) - Z_1^*(t)Y_2(t)$$

is a constant matrix.

In fact if we differentiate this expression and take account of (3) we find that its derivative is 0 . In particular, for any solution Y, Z of (3)

$$Y^*(t)Z(t) - Z^*(t)Y(t)$$

is a constant matrix. The solution Y, Z is said to be *isotropic* if this constant is the zero matrix.

PROPOSITION 1. *Let* Y_0, Z_0 *be an isotropic solution of* (3) *such that* $Y_0(t)$ *is invertible for every* t *in* I . *Then all solutions* Y, Z *of* (3) *are given by the formula*

$$\begin{aligned} Y(t) &= Y_0(t)\left[M+S_0(t)N\right] \\ Z(t) &= Z_0(t)\left[M+S_0(t)N\right] + Y_0^{*-1}(t)N , \end{aligned} \tag{4}$$

where M *and* N *are arbitrary constant matrices and, for some* a *in* I ,

$$S_0(t) = \int_a^t Y_0^{-1}(s)C(s)Y_0^{*-1}(s)ds . \tag{5}$$

Moreover

$$Y_0^*(t)Z(t) - Z_0^*(t)Y(t) = N , \tag{6}$$

$$Y^*(t)Z(t) - Z^*(t)Y(t) = M^*N - N^*M . \tag{7}$$

Proof. It is easily verified by differentiation that the functions Y, Z defined by (4) are solutions of (3). Since

$$Y(a) = Y_0(a)M , \quad Z(a) = Z_0(a)M + Y_0^{*-1}(a)N ,$$

the matrices M, N can be chosen in one and only one way so as to satisfy arbitrary initial conditions. It is sufficient to verify (6) and (7) for $t = a$, since the left sides are independent of t . The relation (7) shows that the solution Y, Z is isotropic if and only if $M^*N = N^*M$.

Suppose the equation (1) is disconjugate on I , *i.e.*, (2) has no nontrivial solution $y(t)$, $z(t)$ such that $y(t_1) = y(t_2) = 0$, where t_1 and t_2 are distinct points of I . Let $Y(t)$, $Z(t)$ be the solution of (3) such that

$$Y(a) = 0 , \quad Z(a) = I ,$$

for some a in I . Then $Y(t)$ is invertible for all $t \neq a$ in I . For suppose $Y(b)\eta = 0$ for some $b \neq a$ in I and some vector η . The solution $y(t) = Y(t)\eta$, $z(t) = Z(t)\eta$ of (2) has $y(a) = y(b) = 0$. It must, therefore, be the trivial solution. In particular, $\eta = z(a) = 0$. The same conclusion is reached if the initial value $Z(a)$ is an arbitrary invertible matrix.

For any symmetric matrix A we write, as usual, $A \geq 0$ if $\eta^*A\eta \geq 0$ for every vector η and $A > 0$ if $\eta^*A\eta > 0$ for every vector $\eta \neq 0$. If A, C are symmetric matrices we write $A \geq C$, $A > C$ if $A-C \geq 0$, resp. $A-C > 0$.

The following result generalises one part of Theorem 1 of Chapter 1.

THEOREM 1. *Suppose* $C(t) \geq 0$ *for all* t *in the compact interval* $I = [a, b]$. *If the equation* (1) *is disconjugate on* I *there exists an isotropic solution* Y, Z *of* (3) *such that* $Y(t)$ *is invertible for all* t *in* I .

Proof. Let Y_1, Z_1 be the solution of (3) such that $Y_1(a) = 0$, $Z_1(a) = I$. Then Y_1, Z_1 is isotropic and, since (1) is disconjugate on I , $Y_1(t)$ is invertible for $a < t \leq b$. Similarly, let Y_2, Z_2 be the solution of (3) such that $Y_2(b) = 0$, $Z_2(b) = -Y_1^{*-1}(b)$. Then Y_2, Z_2 is isotropic, $Y_2(t)$ is invertible for $a \leq t < b$ and

$$Y_2^* Z_1 - Z_2^* Y_1 = I .$$

By Proposition 1 we have $Y_1(t) = Y_2(t)S_2(t)$ for $a \leq t < b$, where

$$S_2(t) = \int_a^t Y_2^{-1}(s)C(s)Y_2^{*-1}(s)ds \geq 0 .$$

Consider the solution $Y = Y_1 + Y_2$, $Z = Z_1 + Z_2$ of (3). Taking $t = b$ we find

$$Y^*Z - Z^*Y = 0 .$$

Thus the solution Y, Z is isotropic. Moreover $Y(b) = Y_1(b)$ is invertible and the relation

$$Y(t) = Y_2(t)\left[I+S_2(t)\right]$$

shows that $Y(t)$ is invertible for $a \leq t < b$.

For convenience of reference we will label the various conditions in which we are interested:

[+] $C(t) \geq 0$ for all t in I ,

[++] $C(t) > 0$ for all t in I ,

[D] the equation (1) is *disconjugate* on I ,

[R] the matrix equation (3) has an isotropic solution Y, Z such that $Y(t)$ is invertible for all t in I .

The labels are intended as mnemonics. It is easily shown (Lemma 7) that condition [R] holds if and only if a corresponding matrix *Riccati* equation has a symmetric solution defined throughout I . We have shown that if I is compact and [+] holds then [D] ⇒ [R]. In order to obtain a converse result we introduce a further *controllability* condition:

[C] if $\Omega(t)$ is a fundamental matrix for the linear equation

$$y' = B(t)y$$

then the rows of $\Omega^{-1}(t)C(t)$ are linearly independent over any subinterval of I .

LEMMA 2. *Condition* [C] *holds if and only if for any nontrivial solution* $y(t)$, $z(t)$ *of* (2) *the component* $y(t)$ *does not vanish throughout any subinterval of* I .

Proof. Suppose $y(t) = 0$ for $a \le t \le b$. Then (2) shows that on the same interval $C(t)z(t) = 0$ and $z(t)$ is a solution of the linear equation

$$z' = - B^*(t)z .$$

Since this equation has the fundamental matrix $\Omega^{*-1}(t)$ we have $z(t) = \Omega^{*-1}(t)\zeta$ for some constant vector ζ . Therefore

$$\zeta^*\Omega^{-1}(t)C(t) = 0 \quad \text{for} \quad a \le t \le b .$$

If [C] holds this implies that $\zeta = 0$. Hence $z(t) = 0$ for $a \le t \le b$ and the original solution is the trivial one. This argument can be reversed.

The meaning of the condition [C] is further brought out by the next result in which the matrix $C(t)$ need not be symmetric or even square.

LEMMA 3. *For two arbitrary vectors* η_1, η_2 *there exists a continuous vector function* $z(t)$ *such that the inhomogeneous linear equation*

$$y' = B(t)y + C(t)z(t) \tag{8}$$

has a solution $y(t)$ *satisfying the boundary conditions*

$$y(t_1) = \eta_1 , \quad y(t_2) = \eta_2$$

if and only if the rows of the matrix $\Omega^{-1}(t)C(t)$ *are linearly independent over the interval* $[t_1, t_2]$.

Proof. Put

$$R = \int_{t_1}^{t_2} \Omega^{-1}(t)C(t)C^*(t)\Omega^{*-1}(t)dt .$$

Evidently R is symmetric and $R \ge 0$. Moreover $\zeta^*R\zeta = 0$ for some vector ζ if and only if

$$\zeta^*\Omega^{-1}(t)C(t) = 0 \quad \text{for} \quad t_1 \le t \le t_2 . \tag{9}$$

Therefore $R > 0$ if and only if the rows of $\Omega^{-1}(t)C(t)$ are linearly independent over $[t_1, t_2]$.

The solutions of (8) are given by

$$y(t) = \Omega(t)\Omega^{-1}(t_1)y(t_1) + \int_{t_1}^{t} \Omega(t)\Omega^{-1}(s)C(s)z(s)ds \ . \tag{10}$$

If we assume $R > 0$ and take

$$z(t) = C^*(t)\Omega^{*-1}(t)R^{-1}\zeta$$

for some constant vector ζ we get

$$y(t_2) = \Omega(t_2)\left[\Omega^{-1}(t_1)y(t_1)+\zeta\right] \ .$$

Hence if we take $y(t_1) = \eta_1$ as initial value and choose

$$\zeta = \Omega^{-1}(t_2)\eta_2 - \Omega^{-1}(t_1)\eta_1$$

then $y(t_2) = \eta_2$.

Conversely, suppose there exists a vector $\zeta \neq 0$ such that (9) holds. From (10) we obtain

$$\Omega(t_1)\Omega^{-1}(t_2)y(t_2) = y(t_1) + \int_{t_1}^{t_2} \Omega(t_1)\Omega^{-1}(s)C(s)z(s)ds \ .$$

Since we can write $\zeta = \Omega^*(t_1)\eta$ for some vector $\eta \neq 0$ it follows that

$$\eta^*\Omega(t_1)\Omega^{-1}(t_2)y(t_2) = \eta^*y(t_1) \ .$$

Hence if $y(t_2) = 0$ then $\eta^*y(t_1) = 0$. Thus $y(t_1) = \eta$ is impossible.

PROPOSITION 2. *Let* Y_0, Z_0 *be an isotropic solution of* (3) *such that* $Y_0(t)$ *is invertible for every* t *in* I . *If* [+] *and* [C] *hold then the symmetric matrix*

$$S_0(t) = \int_a^t Y_0^{-1}(s)C(s)Y_0^{*-1}(s)ds \ ,$$

where $a \in I$, *is an increasing function of* t .

Proof. Evidently [+] implies that $S_0(t)$ is a nondecreasing function of t . Suppose that for some vector η

$$\eta^*\left[S_0(t_2)-S_0(t_1)\right]\eta = 0 \ ,$$

where $t_2 > t_1$. Then for $t_1 \le t \le t_2$

$$\eta^*\left[S_0(t)-S_0(t_1)\right]\eta = 0$$

and hence

$$S_0(t)\eta = S_0(t_1)\eta \; .$$

It follows from Proposition 1 that

$$y(t) = Y_0(t)[-S_0(t_1)+S_0(t)]\eta$$

$$z(t) = Z_0(t)[-S_0(t_1)+S_0(t)]\eta + Y_0^{*-1}(t)\eta$$

is a solution of (2). Since $y(t) = 0$ for $t_1 \le t \le t_2$ it must be the trivial solution, by Lemma 2. Hence $\eta = 0$, which is what we wanted to prove.

We can now obtain a partial converse of Theorem 1.

THEOREM 2. *If conditions* [+] *and* [C] *hold then* [R] $\Rightarrow$ [D].

Proof. Let $y(t)$, $z(t)$ be a solution of (2) such that $y(a) = y(b) = 0$, where $a < b$, and let Y_0, Z_0 be an isotropic solution of (3) such that $Y_0(t)$ is invertible for every t in I. We can suppose this solution normalised so that $Y_0(a) = I$. If Y, Z is the solution of (3) such that $Y(a) = 0$, $Z(a) = I$ then $y(t) = Y(t)\zeta$, $z(t) = Z(t)\zeta$, where $\zeta = z(a)$. By Proposition 1 we have $Y(t) = Y_0(t)S_0(t)$, where

$$S_0(t) = \int_a^t Y_0^{-1}(s)C(s)Y_0^{*-1}(s)ds \; .$$

By Proposition 2, $S_0(t)$ is invertible for $t > a$, and hence $Y(t)$ is also. Since $Y(b)\zeta = 0$ it follows that $\zeta = 0$. Thus $y(t)$, $z(t)$ is the trivial solution and (1) is disconjugate.

2. Principal solutions

We first examine the relationship between two solutions of (3) which have the same properties as Y_0, Z_0 in Proposition 1.

PROPOSITION 3. *Let* Y_1, Z_1 *and* Y_2, Z_2 *be two isotropic solutions of* (3) *such that* $Y_1(t)$ *and* $Y_2(t)$ *are both invertible for every* t *in* I. *If* Y_2, Z_2 *is expressed in terms of* Y_1, Z_1 *by the matrices* M_1, N_1, S_1 *and if* Y_1, Z_1 *is expressed in terms of* Y_2, Z_2 *by the matrices* M_2, N_2, S_2 *then*

$$M_2 = M_1^{-1} ,$$
$$N_2 = - N_1^* , \qquad (11)$$
$$S_2(t) = [M_1+S_1(t)N_1]^{-1}S_1(t)M_1^{*-1} .$$

Proof. Since Y_2, Z_2 is an isotropic solution, (7) implies that

$$M_1^*N_1 = N_1^*M_1 .$$

Moreover, by (6),

$$\begin{aligned} N_2 &= Y_2^*Z_1 - Z_2^*Y_1 \\ &= - [Y_1^*Z_2-Z_1^*Y_2]^* \\ &= - N_1^* , \end{aligned}$$

and by (4),

$$[M_1+S_1(t)N_1][M_2+S_2(t)N_2] = I ,$$

which for $t = a$ gives $M_1M_2 = I$. Thus $M_2 = M_1^{-1}$ and the last identity can be written

$$\begin{aligned} [M_1+S_1(t)N_1]S_2(t)N_2 &= - S_1(t)N_1M_1^{-1} \\ &= S_1(t)M_1^{*-1}N_2 . \end{aligned}$$

If N_1 is invertible the third relation (11) follows immediately.

It remains to prove this relation when N_1 is not invertible. The symmetric matrix $G_1 = M_1^*N_1$ is the limit of a sequence of invertible symmetric matrices $G^{(\nu)}$. If we put

$$N^{(\nu)} = M_1^{*-1}G^{(\nu)}$$

then $N^{(\nu)}$ is invertible, $M_1^*N^{(\nu)} = N^{(\nu)*}M_1$, and $N^{(\nu)} \to N_1$ as $\nu \to \infty$. The functions $Y^{(\nu)}, Z^{(\nu)}$ defined by

$$Y^{(\nu)}(t) = Y_1(t)\left[M_1+S_1(t)N^{(\nu)}\right]$$
$$Z^{(\nu)}(t) = Z_1(t)\left[M_1+S_1(t)N^{(\nu)}\right] + Y_1^{*-1}(t)N^{(\nu)}$$

form an isotropic solution of (3) by Proposition 1. Moreover $Y^{(\nu)}(t) \to Y_2(t)$

uniformly on any compact subinterval J of I as $\nu \to \infty$. Hence $Y^{(\nu)}(t)$ is invertible on J for all large ν. Thus we can write

$$Y_1(t) = Y^{(\nu)}(t)\left[M_1^{-1}-S^{(\nu)}(t)N^{(\nu)*}\right],$$

where $S^{(\nu)}(t) \to S_2(t)$ uniformly on J as $\nu \to \infty$. By what we have already proved

$$S^{(\nu)}(t) = \left[M_1+S_1(t)N^{(\nu)}\right]^{-1}S_1(t)M_1^{*-1}.$$

Letting $\nu \to \infty$ we see that the third relation (11) continues to hold.

COROLLARY. *If* $S_1(t)$ *is invertible then* $S_2(t)$ *is also invertible and*

$$S_2^{-1}(t) = M_1^*\left[S_1^{-1}(t)M_1+N_1\right]. \tag{12}$$

From now until further notice *we assume that* $I = [a, b)$ *is a half-open interval and that conditions* [+], [R] *and* [C] *hold.*

Thus the equation (3) has an isotropic solution Y_0, Z_0 such that $Y_0(t)$ is invertible for all t in I. By Proposition 2 the symmetric matrix

$$S_0(t) = \int_a^t Y_0^{-1}(s)C(s)Y_0^{*-1}(s)ds$$

is an increasing function of t and hence invertible for $t > a$. It follows that $S_0^{-1}(t)$ is a decreasing function of t, on account of

LEMMA 4. *If* $A > B > 0$ *then* $B^{-1} > A^{-1}$.

Proof. Suppose first that $A = I$. From $0 < B < I$ we get

$$B^{-1} = [I-(I-B)]^{-1} = I + \sum_{n=1}^{\infty} (I-B)^n > I.$$

In the general case there exists a matrix $C > 0$ such that $C^2 = A$. From $0 < B < A$ we get $0 < C^{-1}BC^{-1} < I$. Therefore, by what we have already proved, $CB^{-1}C > I$ and hence $B^{-1} > C^{-2} = A^{-1}$.

Consequently there exists a symmetric matrix $T_0 \geq 0$ such that

$$S_0^{-1}(t) \to T_0 \quad \text{as} \quad t \to b.$$

Put

$$\hat{Y}(t) = Y_0(t)[I - S_0(t)T_0]$$

$$\hat{Z}(t) = Z_0(t)[I - S_0(t)T_0] - Y_0^{*-1}(t)T_0 .$$

Then $\hat{Y}$, $\hat{Z}$ is an isotropic solution of (3), by Proposition 1. Moreover $\hat{Y}(t)$ is invertible for every t in I, since $\hat{Y}(a) = Y_0(a)$ and for $t > a$

$$\hat{Y}(t) = Y_0(t)S_0(t)\left[S_0^{-1}(t) - T_0\right]$$

is a product of invertible matrices. Moreover, by (12)

$$\hat{S}^{-1}(t) \equiv \left[\int_a^t \hat{Y}^{-1}(s)C(s)\hat{Y}^{*-1}(s)ds\right]^{-1} .$$

$$= S_0^{-1}(t) - T_0 ,$$

and hence $\hat{S}^{-1}(t) \to 0$ as $t \to b$. For any constant matrix K

$$[\hat{S}(t)+K]^{-1} = \hat{S}^{-1}(t)[I + K\hat{S}^{-1}(t)]^{-1} \to 0 \quad \text{as} \quad t \to b .$$

This shows that the property $\hat{S}^{-1}(t) \to 0$ still holds if we change the lower limit of integration in $\hat{S}(t)$ from a to c.

Suppose now that (3) has an isotropic solution Y, Z such that $Y(t)$ is invertible on an interval $[c, b)$. Then as $t \to b$,

$$S^{-1}(t) \equiv \left[\int_c^t Y^{-1}(s)C(s)Y^{*-1}(s)ds\right]^{-1} \to T , \text{ say.}$$

We can write

$$Y(t) = \hat{Y}(t)[M + \hat{S}(t)N]$$

$$Z(t) = \hat{Z}(t)[M + \hat{S}(t)N] + \hat{Y}^{*-1}N ,$$

where now

$$\hat{S}(t) = \int_c^t \hat{Y}^{-1}(s)C(s)\hat{Y}^{*-1}(s)ds .$$

By (12) we have

$$S^{-1}(t) = M^*[\hat{S}^{-1}(t)M + N] .$$

Letting $t \to b$ we get $T = M^*N$. Since M is invertible, $T = 0$ if and only if $N = 0$, *i.e.*, if and only if

$$Y(t) = \hat{Y}(t)M , \quad Z(t) = \hat{Z}(t)M .$$

It then follows that $Y(t)$ is invertible for every t in I.

We define a solution Y, Z *of* (3) *to be a principal solution if it is isotropic, if* $Y(t)$ *is invertible on an interval* $[c, b)$ *and if*

$$\left[\int_c^t Y^{-1}(s)C(s)Y^{*-1}(s)ds\right]^{-1} \to 0 \quad as \quad t \to b \; .$$

Then the preceding argument is summed up in

THEOREM 3. *A principal solution* Y, Z *exists, it is uniquely determined up to a constant invertible factor, and* $Y(t)$ *is invertible for every* t *in* I .

Our next result gives another characterization of the principal solution.

PROPOSITION 4. *Let* $\hat{Y}, \hat{Z}$ *be an isotropic solution of* (3) *such that* $\hat{Y}(t)$ *is invertible for all* t *near* b *and let* Y, Z *be any solution of* (3). *Then* $\hat{Y}, \hat{Z}$ *is a principal solution and the constant matrix*

$$N = \hat{Y}^*Z - \hat{Z}^*Y$$

is invertible if and only if $Y(t)$ *is invertible for all* t *near* b *and* $Y^{-1}(t)\hat{Y}(t) \to 0$ *as* $t \to b$.

Proof. By Proposition 1, for any c near b we have

$$Y(t) = \hat{Y}(t)[M+\hat{S}(t)N] \; ,$$

where

$$\hat{S}(t) = \int_c^t \hat{Y}^{-1}(s)C(s)\hat{Y}^{*-1}(s)ds \; .$$

Suppose first that $\hat{Y}, \hat{Z}$ is a principal solution and N is invertible. Then $\hat{S}^{-1}(t) \to 0$ as $t \to b$ and hence

$$M + \hat{S}(t)N = \hat{S}(t)\left[\hat{S}^{-1}(t)M+N\right]$$

is invertible for all t near b . Therefore $Y(t)$ is invertible and

$$Y^{-1}(t)\hat{Y}(t) = \left[\hat{S}^{-1}(t)M+N\right]^{-1}\hat{S}^{-1}(t) \to 0 \quad \text{as} \quad t \to b \; .$$

Conversely, suppose that $Y(t)$ is invertible for all t near b and $Y^{-1}(t)\hat{Y}(t) \to 0$ as $t \to b$. Then $M = \hat{Y}^{-1}(c)Y(c)$ is invertible if c is sufficiently near b and

$$\left[I+\hat{S}(t)NM^{-1}\right]^{-1} \to 0 \quad \text{as} \quad t \to b \; .$$

Hence if $0 < \varepsilon < 1$ we will have for $t \geq t_\varepsilon$ and for every vector η

$$\begin{aligned} |\eta| &\leq \varepsilon\left|\left[I+\hat{S}(t)NM^{-1}\right]\eta\right| \\ &\leq \varepsilon|\eta| + \varepsilon|\hat{S}(t)NM^{-1}\eta| \; . \end{aligned}$$

Therefore

$$|\eta| \le \varepsilon(1-\varepsilon)^{-1}|\hat{S}(t)NM^{-1}\eta| ,$$

which shows that N is invertible and

$$|MN^{-1}\hat{S}^{-1}(t)| \le \varepsilon(1-\varepsilon)^{-1} \quad \text{for} \quad t \ge t_\varepsilon .$$

Consequently

$$|\hat{S}^{-1}(t)| \le |NM^{-1}||MN^{-1}\hat{S}^{-1}(t)| \to 0 \quad \text{as} \quad t \to b .$$

Under the present hypotheses the equation (1) is disconjugate on I, by Theorem 2. Hence for any $s > a$ the system (3) has a unique solution Y_s, Z_s such that

$$Y_s(a) = I , \quad Y_s(s) = 0 .$$

The boundary condition at s implies that Y_s, Z_s is an isotropic solution. Let Y_0, Z_0 be an isotropic solution of (3) such that $Y_0(t)$ is invertible for all t in I. We can suppose this solution normalised so that $Y_0(a) = I$. If M_s, N_s are the matrices representing Y_s, Z_s in terms of Y_0, Z_0 according to Proposition 1 then the boundary conditions at a and s give

$$M_s = I , \quad N_s = - S_0^{-1}(s) .$$

Thus N_s is an increasing symmetric matrix function. Since $N_s \le 0$ for all s it follows that the limit

$$N_b = \lim_{s \to b} N_s$$

exists and $N_b \le 0$. Consider the solution

$$Y_b(t) = Y_0(t)[I+S_0(t)N_b]$$

$$Z_b(t) = Z_0(t)[I+S_0(t)N_b] + Y_0^{*-1}(t)N_b .$$

Evidently $Y_s(t)$, $Z_s(t) \to Y_b(t)$, $Z_b(t)$ uniformly on compact subintervals of I as $s \to b$ and the solution Y_b, Z_b is isotropic. Moreover $Y_b(t)$ is invertible for all t in I. This is obvious for $t = a$ and for $a < t < b$ it follows from the representation

$$Y_b(t) = Y_0(t)S_0(t)\left[S_0^{-1}(t)+N_b\right] .$$

If we represent Y_0, Z_0 in terms of Y_b, Z_b then, by (12),

$$S_b^{-1}(t) \equiv \left[\int_a^t Y_b^{-1}(s)C(s)Y_b^{*-1}(s)ds\right]^{-1}$$

$$= S_0^{-1}(t) + N_b \to 0 \quad \text{as} \quad t \to b .$$

Therefore Y_b, Z_b is a principal solution. Thus *a principal solution can be constructed as the limit of the solutions* Y_s, Z_s .

Finally we use the properties of principal solutions to derive some results for open intervals.

THEOREM 4. *Suppose conditions* [+] *and* [C] *hold on an open interval* $I = (a, b)$. *Then* [D] $\Leftrightarrow$ [R].

Proof. By Theorem 2, [R] $\Rightarrow$ [D]. Thus we need only prove [D] $\Rightarrow$ [R]. For any c in I let Y_c, Z_c be the solution of (3) such that $Y_c(c) = 0$, $Z_c(c) = I$. Then Y_c, Z_c is isotropic and $Y_c(t)$ is invertible for $t \neq c$. Hence for any d such that $c < d < b$ the system (3) has a principal solution $\hat{Y}$, $\hat{Z}$ on the interval $[d, b)$. In particular $\hat{Y}$, $\hat{Z}$ is an isotropic solution. Since the principal solution is uniquely determined up to a constant invertible factor and d can be any point of I , it follows that $\hat{Y}(t)$ is invertible for every t in I .

THEOREM 5. *Suppose* [+] *and* [C] *hold on the half-open interval* $I = [a, b)$. *Then* (1) *is disconjugate on* I *if, and only if, it is disconjugate on its interior* (a, b) .

Proof. By Theorem 4 the system (3) has an isotropic solution Y_0, Z_0 such that $Y_0(t)$ is invertible for all t in (a, b) . Hence for any c , $a < c < b$, the system (3) has a principal solution $\hat{Y}$, $\hat{Z}$ on the half-open interval $(a, c]$. If Y, Z is the solution of (3) such that $Y(c) = 0$, $Z(c) = I$ then $\hat{Y}^*Z - \hat{Z}^*Y$ is invertible. Therefore, by Proposition 4, $Y(t)$ is invertible for all t near a and

$$Y^{-1}(t)\hat{Y}(t) \to 0 \quad \text{as} \quad t \to a .$$

Since $Y(t) \to Y(a)$ as $t \to a$ it follows that $\hat{Y}(t) \to 0$. Thus $\hat{Y}(a) = 0$, and hence $\hat{Z}(a)$ is invertible. Therefore any solution $y(t)$, $z(t)$ of (2) such that $y(a) = 0$ has the form

$$y(t) = \hat{Y}(t)\eta , \quad z(t) = \hat{Z}(t)\eta ,$$

for some constant vector η . Since $\hat{Y}(c)$ is invertible, $y(c) = 0$ implies $\eta = 0$. Therefore, for no nontrivial solution $y(t)$, $z(t)$, do we have $y(a) = y(c) = 0$, $(a < c < b)$. This is all that requires proof.

3. Conjugate points

We suppose throughout this section that I *is an open interval and conditions* [+] *and* [C] *hold.* For any a in I there exists $\delta = \delta(a) > 0$ such that the equation (1) is disconjugate on the subinterval $[a-\delta, a+\delta]$. This follows from Theorem 2, since if Y, Z is the solution of (3) such that $Y(a) = I$, $Z(a) = 0$, then $Y(t)$ is invertible in some neighbourhood of a.

Two points a, b of I are said to be *conjugate* if there exists a nontrivial solution $y(t), z(t)$ of (2) such that $y(a) = y(b) = 0$. Let $Y(t, a), Z(t, a)$ be the isotropic solution of (3) which satisfies the initial condition $Y = 0$, $Z = I$ at $t = a$. Then $Y(t, s), Z(t, s)$ are continuous functions on $I \times I$.

LEMMA 5. *Two points* a, b *of* I *are conjugate if and only if* $\det Y(b, a) = 0$.

Proof. Any nontrivial solution $y(t), z(t)$ of (2) such that $y(a) = 0$ has the form $y(t) = Y(t, a)\eta$, $z(t) = Z(t, a)\eta$ for some vector $\eta \neq 0$. We can choose η so that $y(b) = 0$ if and only if $\det Y(b, a) = 0$.

Suppose now that the equation (1) is *not* disconjugate on I. Then for some a in I there exists $b > a$ in I such that (1) is not disconjugate on $[a, b]$. Moreover, if a has this property, so also does any $a' < a$ in I. We denote by $\sigma_+(a) = \sigma(a)$ the supremum of all $c > a$ such that (1) is disconjugate on $[a, c]$. Thus if $\sigma(a)$ is defined, so is $\sigma(a')$ for any $a' < a$ in I and $\sigma(a') \leq \sigma(a)$. Also let $\omega_+(a) = \omega(a)$ be the least $b > a$ in I, if one exists, such that $\det Y(b, a) = 0$.

PROPOSITION 5. $\sigma(a) = \omega(a)$.

Proof. By Lemma 5 and the definition of $\sigma(a)$ we have $\sigma(a) \leq \omega(a)$. Suppose $\sigma(a) < \omega(a)$ and choose c in I so that $\sigma(a) < c < \omega(a)$. The equation (1) is disconjugate on an interval $[a-\delta, a+\delta]$, where $\delta > 0$. Since $Y(t, s)$ is continuous and $\det Y(t, a) \neq 0$ for $a < t < \omega(a)$ we can choose $\delta_0 < \delta/2$ so that $\det Y(t, s) \neq 0$ for $a-\delta_0 \leq s \leq a+\delta_0$, $a+\delta/2 \leq t \leq c$. Since (1) is disconjugate on $[a-\delta, a+\delta]$ we have $\det Y(t, a-\delta_0) \neq 0$ for $a-\delta_0 < t < a+\delta/2$. Thus $\det Y(t, a-\delta_0) \neq 0$ for $a-\delta_0 < t \leq c$. Therefore, by Theorem 2, the equation (1) is disconjugate on $[a-\delta_0, c]$. But this implies $\sigma(a) \geq c$, which is a contradiction.

THEOREM 6. $\sigma(a)$ *is an increasing function.*

Proof. We already know that $\sigma(a)$ is a nondecreasing function. Suppose $a_1 < a_2$ and $\sigma(a_1) = \sigma(a_2) = b$. Then $\sigma(t) = b$ for $a_1 \leq t \leq a_2$. Therefore, by Proposition 5 and Lemma 5, t and b are conjugate points and $\det Y(t, b) = 0$ for

$a_1 \leq t \leq a_2$. Let $n - p$ $(p \geq 1)$ be the maximum rank of $Y_2(t) = Y(t, b)$ in the interval $[a_1, a_2]$. Then there exists a subinterval J on which an $(n-p) \times (n-p)$ minor of $Y_2(t)$ never vanishes. The equations

$$\eta^* Y_2(t) = 0 \tag{13}$$

then have p linearly independent solutions $\eta_1(t), \ldots, \eta_p(t)$, which we can take as continuously differentiable functions, on J .

If $Y_1(t)$, $Z_1(t)$ is the isotropic solution of (3) such that $Y_1(b) = I$, $Z_1(b) = 0$ then

$$Y_1^* Z_2 - Z_1^* Y_2 = I .$$

It follows that the matrix

$$\begin{pmatrix} Z_2^* & -Y_2^* \\ -Z_1^* & Y_1^* \end{pmatrix}$$

is a left inverse for the matrix

$$\begin{pmatrix} Y_1 & Y_2 \\ Z_1 & Z_2 \end{pmatrix}$$

and hence also a right inverse. Thus

$$Y_2 Y_1^* - Y_1 Y_2^* = 0 ,$$

$$Z_2 Y_1^* - Z_1 Y_2^* = I .$$

It follows that

$$Y_2' Y_1^* - Y_1' Y_2^* = C\left(Z_2 Y_1^* - Z_1 Y_2^*\right) = C .$$

Thus

$$C = Y_1 Y_2^{*\prime} - Y_2 Y_1^{*\prime} .$$

Hence for any continuously differentiable solution $\eta(t)$ of (13)

$$\eta^* C \eta = \eta^* Y_1 Y_2^{*\prime} \eta .$$

But

$$0 = \left(Y_2^* \eta\right)' = Y_2^{*\prime} \eta + Y_2^* \eta' .$$

Therefore

$$\eta^* C\eta = -\eta^* Y_1 Y_2^* \eta'$$
$$= -\eta^* Y_2 Y_1^* \eta'$$
$$= 0 .$$

Hence $C\eta = 0$. Since $Y_2' = BY_2 + CZ_2$ it follows that $\eta^* Y_2' = \eta^* BY_2$, and therefore

$$Y_2^*(\eta' + B^*\eta) = \left(-Y_2^{*\prime} + Y_2^* B^*\right)\eta = 0 .$$

Consequently we can write

$$\eta_j' + B^*\eta_j = \sum_{k=1}^{p} \omega_{jk}\eta_k \quad (j = 1, \ldots, p)$$

with continuous scalar coefficients $\omega_{jk}(t)$. Choose scalar functions $\lambda_j(t)$, not all zero, such that

$$\lambda_j' + \sum_{k=1}^{p} \omega_{kj}\lambda_k = 0 \quad (j = 1, \ldots, p) .$$

Then $\eta(t) = \sum_{j=1}^{p} \lambda_j \eta_j$ is a nontrivial solution of (13) and

$$\eta' + B^*\eta = 0 .$$

If $\Omega(t)$ is a fundamental matrix for the equation $y' = By$ it follows that $\eta(t) = \Omega^{*-1}(t)\zeta$ on J, for some constant vector $\zeta \neq 0$. However, since $\eta^* C = 0$, this contradicts condition [C].

Theorem 6 can also be deduced very simply from Theorem 5. Suppose $a_1 < a_2$ and $\sigma(a_1) = \sigma(a_2) = b$. Choose c so that $a_1 < c < a_2$. The equation (1) is disconjugate on the interval $[a_1, b)$, and hence on (c, b). Therefore, by Theorem 5, it is disconjugate on $(c, b]$, and hence on $[a_2, b]$. But this contradicts Proposition 5.

Similarly for some b in I there exists $a < b$ in I such that (1) is not disconjugate on $[a, b]$. Moreover if b has this property so also does any $b' > b$ in I. We denote by $\sigma_-(b)$ the infimum of all $c < b$ such that (1) is disconjugate on $[c, b]$. Thus if $\sigma_-(b)$ is defined, so is $\sigma_-(b')$ for any $b' > b$ in I and $\sigma_-(b') \geq \sigma_-(b)$. In the same way as for σ_+ we can show that σ_- is an increasing function.

LEMMA 6. *If* $b = \sigma_+(a)$ *then* $a = \sigma_-(b)$.

Proof. By Lemma 5 and Proposition 5 the points a and b are conjugate.

Therefore $c = \sigma_-(b) \geq a$. For the same reason b and c are conjugate and hence $\sigma_+(c) \leq b$. Since $c > a$ would imply $\sigma_+(c) > \sigma_+(a) = b$, we conclude that $c = a$.

It follows that the functions σ_+ and σ_- are inverses of one another. Since their domains of definition are intervals, their ranges are also intervals. Therefore, since they are increasing functions, they are continuous. Suppose $I = (a_1, b_1)$. If the domain of σ_+ were a half-closed interval $(a_1, a]$ then any $b > \sigma_+(a)$ in I would not belong to the range of σ_+ and therefore would not belong to the domain of σ_- . But this is impossible because the domain of σ_- is an interval with endpoint b_1 . Hence we have proved

THEOREM 7. *The function* $\sigma(a)$ *is continuous and its domain is an open subinterval of* I .

As an application we give another proof of Theorem 5. Suppose [+] and [C] hold on the half-open interval $[a, b)$. If we extend the domain of definition of the coefficient matrix in (1) by setting $H(t) = H(a)$ for $t < a$, then the preceding results can be applied to the open interval $\tilde{I} = (-\infty, b)$. If (1) were disconjugate on (a, b) but not on $[a, b)$ then $\sigma(a)$ would be defined, but not $\sigma(c)$ for any $c > a$. However this contradicts Theorem 7.

4. Matrix Riccati equations

The connection between linear Hamiltonian systems and matrix Riccati equations is described in the following lemma, whose proof is immediate.

LEMMA 7. *If* $Y(t)$, $Z(t)$ *is a solution of* (3) *such that* $Y(t)$ *is invertible for all* t *in* I *then* $W(t) = Z(t)Y^{-1}(t)$ *is a solution of the equation*

$$R[W] \equiv W' + A(t) + WB(t) + B^*(t)W + WC(t)W = 0 . \qquad (14)$$

The solution $Y(t)$, $Z(t)$ *of* (3) *is isotropic if and only if the corresponding solution* $W(t)$ *of* (14) *is symmetric.*

Conversely, if $W(t)$ *is a solution of* (14) *and if* $Y(t)$ *is a fundamental solution of the linear equation*

$$y' = [B(t)+C(t)W(t)]y \qquad (15)$$

then $Y(t)$ *and* $Z(t) = W(t)Y(t)$ *are a solution of* (3).

This shows that condition [R] is equivalent to the existence throughout I of a symmetric solution of (14). If $W(t)$ is a solution of (14) then $W^*(t)$ is also a solution. Since the solutions of (14) are uniquely determined by their initial values it follows that a solution $W(t)$ is symmetric throughout its interval of definition (which may be smaller than the interval I throughout which the coefficients A, B, C are continuous) if it is symmetric at one point.

The matrix Riccati equation also solves the problem of factorisation of a linear Hamiltonian equation.

LEMMA 8. *Suppose* (14) *has a symmetric solution* $W(t)$ *on the interval* I . *If* $y(t)$, $z(t)$ *are continuously differentiable functions such that*

$$y' = B(t)y + C(t)z$$

then

$$z' + A(t)y + B^*(t)z = L_1^*[z-Wy] , \tag{16}$$

where

$$L_1^*[u] = u' + (B^*+WC)u .$$

If $C(t)$ *is invertible on* I *this can be written in the form*

$$z' + A(t)y + B^*(t)z = L_1^*\left[C^{-1}L_1(y)\right] ,$$

where

$$L_1[u] = u' - (B+CW)u .$$

We show next that the basic results on differential inequalities for scalar equations can be extended to matrix Riccati equations.

PROPOSITION 6. *Let* $W_1(t)$, $W_2(t)$ *be symmetric solutions of the Riccati equation* (14) *on an interval* I . *If, for some* a *in* I , $W_2(a) \geq W_1(a)$ *then* $W_2(t) \geq W_1(t)$ *for all* t *in* I . *If* $W_2(a) > W_1(a)$ *then* $W_2(t) > W_1(t)$ *for all* t *in* I .

Proof. The difference $V(t) = W_2(t) - W_1(t)$ is a solution of the equation

$$V' + V(B+CU) + (B^*+UC)V = 0 ,$$

where $U(t) = \left[W_1(t)+W_2(t)\right]/2$. If $\Phi(t)$ is the fundamental matrix for the equation

$$y' = (B+CU)y$$

such that $\Phi(a) = I$ it follows that

$$V(t) = \Phi^{*-1}(t)V(a)\Phi^{-1}(t) .$$

This formula shows that $V(t) \geq 0$ if $V(a) \geq 0$ and the nullspace of $V(t)$ has the same dimension as the nullspace of $V(a)$.

PROPOSITION 7. *Let* $W(t)$ *be a symmetric solution of the Riccati equation* (14) *on the interval* $I = [a, b]$.

If $U(t)$ *is a symmetric solution of the inequality* $R[U] > 0$ *on* I *such that* $U(a) > W(a)$, *then* $U(t) > W(t)$ *throughout* I .

If $V(t)$ *is a symmetric solution of the inequality* $R[V] < 0$ *on* I *such that* $V(a) < W(a)$, *then* $V(t) < W(t)$ *throughout* I .

Proof. If the inequality $U(t) > W(t)$ does not hold throughout I there is a least value $c > a$ for which it fails to hold. Then $U(t) > W(t)$ for $a \le t < c$, $U(c) \ge W(c)$, and $U(c)\xi = W(c)\xi$ for some vector $\xi \ne 0$. Put

$$\varphi(t) = \xi^*[U(t)-W(t)]\xi .$$

Then $\varphi(t) > 0$ for $a \le t < c$ and $\varphi(c) = 0$. But

$$\begin{aligned}\varphi'(c) &= \xi^*U'(c)\xi - \xi^*W'(c)\xi \\ &= \xi^*R[U](c)\xi - \xi^*R[W](c)\xi \\ &> 0 .\end{aligned}$$

Therefore $\varphi(t) < 0$ if t is less than c and sufficiently close to c , which is a contradiction. The other half of the proposition is proved similarly.

PROPOSITION 8. *Let* $W(t)$ *be a symmetric solution of the Riccati equation* (14) *on the interval* $I = [a, b]$.

If $U(t)$ *is a symmetric solution of the inequality* $R[U] \ge 0$ *on* I *such that* $U(a) \ge W(a)$, *then* $U(t) \ge W(t)$ *throughout* I .

If $V(t)$ *is a symmetric solution of the inequality* $R[V] \le 0$ *on* I *such that* $V(a) \le W(a)$, *then* $V(t) \le W(t)$ *throughout* I .

Proof. If the inequality $U(t) \ge W(t)$ does not hold throughout I there is a greatest value $c < b$ such that it holds throughout $[a, c]$. Let $W^{(\nu)}(t)$ be the solution of the equation

$$R[W] + I/\nu = 0$$

which satisfies the initial condition

$$W^{(\nu)}(c) = W(c) - I/\nu .$$

Then $W^{(\nu)}(t)$ is defined throughout an interval $[c, c+\delta]$, where $\delta > 0$ is independent of ν . By Proposition 7

$$U(t) > W^{(\nu)}(t) \quad \text{for} \quad c \le t \le c+\delta .$$

Since $W^{(\nu)}(t) \to W(t)$ as $\nu \to \infty$ it follows that

$$U(t) \ge W(t) \quad \text{for} \quad c \le t \le c+\delta ,$$

which contradicts the definition of c. The second part of the proposition is proved similarly.

PROPOSITION 9. *Let* $U(t)$, $V(t)$ *be symmetric solutions of the inequalities* $R[U] \geq 0$, $R[V] \leq 0$ *respectively on the interval* $I = [a, b]$, *and suppose* $U(t) \geq V(t)$ *throughout* I.

Then for any symmetric matrix $\tilde{W}$ *such that* $U(a) \geq \tilde{W} \geq V(a)$ *the solution* $W(t)$ *of the Riccati equation* (14) *which satisfies the initial condition* $W(a) = \tilde{W}$ *is defined and satisfies the inequalities* $U(t) \geq W(t) \geq V(t)$ *throughout* I.

Proof. If the solution $W(t)$ is not continuable throughout I then $|W(t)| \to \infty$ as $t \uparrow c$ for some $c > a$. But by Proposition 8 the inequalities $U(t) \geq W(t) \geq V(t)$ hold for $a \leq t < c$ and hence also for $t = c$.

Proposition 9 holds equally well if $I = [a, b)$ is a half-open interval.

COROLLARIES.

(i) If in Proposition 8 *we have strict inequality in the initial condition then we also have strict inequality in the conclusion.*

Suppose, for example, that $U(a) > W(a)$. The solution $\tilde{W}(t)$ of (14) such that $\tilde{W}(a) = U(a)$ is defined and satisfies $U(t) \geq \tilde{W}(t) \geq W(t)$ throughout I. By Proposition 6 we must actually have $\tilde{W}(t) > W(t)$, and hence $U(t) > W(t)$.

(ii) Suppose $C(t) \geq 0$ *and the inequality* $R[V] \leq 0$ *has a symmetric solution* $V(t)$ *defined on* $I = [a, b]$ *or* $[a, b)$. *If* $W(t)$ *is a symmetric solution of the Riccati equation* (14) *such that* $W(a) \geq V(a)$ *then* $W(t)$ *is defined and satisfies the inequality* $W(t) \geq V(t)$ *throughout* I.

For any $P > 0$ the solution $U(t)$ of the linear equation

$$U' + A(t) - P + UB(t) + B^*(t)U = 0$$

such that $U(a) = W(a) + P$ is defined throughout I. By Proposition 7 we have $U(t) > V(t)$ throughout I. Therefore, by Proposition 9, $W(t)$ is defined and satisfies $U(t) \geq W(t) \geq V(t)$ for all t in I.

(iii) Suppose $C(t) \geq 0$ *and* $A(t) \leq 0$ *on the interval* $I = [a, b]$ *or* $[a, b)$. *If* $W(t)$ *is a solution of the Riccati equation* (14) *such that* $W(a) \geq 0$ *then it is defined and satisfies the inequality* $W(t) \geq 0$ *throughout* I.

This is a special case of *(ii)*, since $R[0] = A(t) \leq 0$.

(iv) Suppose we have two linear Hamiltonian systems

$$Jx' = H_1(t)x \tag*{$(1)_1$}$$

$$Jx' = H_2(t)x, \tag*{$(1)_2$}$$

where $H_1(t) \leq H_2(t)$ *and* $(1)_1$ *satisfies condition* [+] *on the interval* $I = [a, b]$

or $[a, b)$. *Let* $W_2(t)$ *be a symmetric solution of the Riccati equation* $R_2[W] = 0$ *on* I . *If* $W_1(t)$ *is a symmetric solution of the Riccati equation* $R_1[W] = 0$ *such that* $W_1(a) \geq W_2(a)$ *then* $W_1(t)$ *is defined and satisfies the inequality* $W_1(t) \geq W_2(t)$ *throughout* I .

This follows from *(ii)*, since

$$R[W] = W' + (I \ W)H\begin{pmatrix} I \\ W \end{pmatrix}$$

and hence $R_1[W_2] \leq R_2[W_2] = 0$.

From this last result it is easy to obtain an extension to linear Hamiltonian equations of Sturm's comparison theorem.

PROPOSITION 10. *Suppose we have two linear Hamiltonian systems* $(1)_1$ *and* $(1)_2$, *where* $H_1(t) \leq H_2(t)$ *for all* t *in an interval* I . *Suppose also that the first equation satisfies conditions* [+] *and* [C]. *Then if the equation* $(1)_2$ *is disconjugate on* I *the equation* $(1)_1$ *is also.*

Proof. Let J be any compact subinterval of I . Since $C_2 \geq C_1 \geq 0$, the equation $(1)_2$ satisfies [R] on J , by Theorem 1. Therefore, by Corollary *(iv)* above, the equation $(1)_1$ also satisfies [R]. By Theorem 2 this implies that it is disconjugate on J , and hence on I .

Suppose now that conditions [+], [R] and [C] hold on the half-open interval $I = [a, b)$, so that the system (3) has a principal solution. Since the principal solution $\hat{Y}$, $\hat{Z}$ of (3) is uniquely determined up to a constant invertible factor, $\hat{W} = \hat{Z}\hat{Y}^{-1}$ is a *uniquely* determined symmetric solution of the corresponding Riccati equation (14). We will call it the principal solution of (14). We show next how $\hat{W}$ can be characterized as a solution of (14) without reference to (3). By means of Lemma 7 we first reformulate Proposition 1 in the following way:

Let $W_1(t)$ be a symmetric solution of (14) for $t \in I$. For some fixed $a \in I$ let $Y_1(t)$ be the fundamental solution of the linear equation

$$y' = [B(t)+C(t)W_1(t)]y$$

such that $Y_1(a) = I$ and set

$$S_1(t) = \int_a^t Y_1^{-1}(s)C(s)Y_1^{*-1}(s)ds .$$

Then $W_2(t)$ is also a solution of (14) for $t \in I$ if and only if $N = W_2(a) - W_1(a)$ is such that $I + S_1(t)N$ is invertible for $t \in I$, in which case

$$W_2(t) - W_1(t) = Y_1^{*-1}(t)N[I+S_1(t)N]^{-1}Y_1^{-1}(t) .$$

If N is symmetric then in the same way we can represent $W_1(t)$ in terms of $W_2(t)$. Moreover, by (12), if $S_1(t)$ is invertible then $S_2(t)$ is also invertible and

$$S_2^{-1}(t) = S_1^{-1}(t) + N .$$

THEOREM 8. *Suppose conditions* [+], [R] *and* [C] *hold on the half-open interval* $I = [a, b)$. *The principal solution* $\hat{W}(t)$ *of the Riccati equation* (14) *is a symmetric solution defined throughout* I *with the property that an arbitrary symmetric solution* $W(t)$ *is defined throughout* I *if and only if* $W(a) \geq \hat{W}(a)$. *Moreover we then have* $W(t) \geq \hat{W}(t)$ *for all* t *in* I.

Proof. Put $N = W(a) - \hat{W}(a)$. By the reformulated version of Proposition 1, $W(t)$ is defined throughout I if and only if $I + \hat{S}(t)N$ is invertible for all t in I, *i.e.*, if and only if $\hat{S}^{-1}(t) + N$ is invertible for $a < t < b$. Since $\hat{S}^{-1}(t) > 0$ it is clear that $N \geq 0$ is a sufficient condition. On the other hand, if $W(t)$ is defined throughout I, we can represent $\hat{W}$ in terms of W with

$$S^{-1}(t) = \hat{S}^{-1}(t) + N .$$

Thus $S^{-1}(t) \to N$ as $t \to b$. Since $S^{-1}(t) > 0$ it follows that $N \geq 0$. This proves the first statement of the theorem, and the second follows from Proposition 6.

Thus the principal solution of (14) is the minimal symmetric solution in the neighbourhood of b. Theorem 8 has the following

COROLLARY. *Suppose we have two linear Hamiltonian systems* $(1)_1$ *and* $(1)_2$, *with* $H_1(t) \leq H_2(t)$, *satisfying conditions* [+], [R] *and* [C] *on the interval* $I = [a, b)$. *Then the principal solutions of the corresponding Riccati equations satisfy* $\hat{W}_1(t) \leq \hat{W}_2(t)$ *throughout* I.

In fact for any c in I the solution $W_1(t)$ of the Riccati equation $R_1[W] = 0$ such that $W_1(c) = \hat{W}_2(c)$ is defined and satisfies $W_1(t) \geq \hat{W}_2(t)$, for $c \leq t < b$, by Corollary *(iv)* to Proposition 9. Since $\hat{W}_1(t)$ is a principal solution of the equation $R_1[W] = 0$ also on the interval $[c, b)$ it follows that $W_1(t) \geq \hat{W}_1(t)$ for $c \leq t < b$. In particular $\hat{W}_2(c) \geq \hat{W}_1(c)$.

Similarly, if the conditions of Corollary *(iii)* of Proposition 9 are satisfied and in addition [C] holds then for the principal solution $\hat{W}(t)$ of (14) we have $\hat{W}(t) \leq 0$ throughout I.

THEOREM 9. *Suppose conditions* [+], [R] *and* [C] *hold on the open interval* $I = (a, b)$. *The right and left principal solutions* $W_+(t)$, $W_-(t)$ *of the Riccati equation* (14) *are symmetric solutions defined throughout* I *with the property that, for any* c *in* I, *an arbitrary symmetric solution* $W(t)$ *is defined throughout* I *if and only if* $W_-(c) \geq W(c) \geq W_+(c)$. *Moreover we then have* $W_-(t) \geq W(t) \geq W_+(t)$ *for all* t *in* I.

This follows from Theorem 8 and Proposition 6, and their analogues with t replaced by $-t$.

We next apply these results to the autonomous linear Hamiltonian equation

$$Jx' = Hx \,, \tag{17}$$

where

$$H = \begin{pmatrix} A & B^* \\ B & C \end{pmatrix}$$

is a constant symmetric matrix. If $W(t)$ is a solution of the corresponding Riccati equation

$$W' + A + WB + B^*W + WCW = 0 \tag{18}$$

on the interval $[a, b]$ then $W(t-h)$ is a solution on the interval $[a+h, b+h]$. It follows, that, if conditions [+] and [C] hold, then (17) is disconjugate on a half-line $[a, \infty)$ only if it is disconjugate on the whole line $(-\infty, \infty)$. If $W_+(t)$ is the principal solution of (18) on the interval $[a, \infty)$ then $W_+(t+h)$ is a solution on the interval $[a-h, \infty)$. Hence $W_+(t+h) \geq W_+(t)$ for $t \geq a$, and in particular $W_+(a+h) \geq W_+(a)$. Similarly, $W_+(t-h) \geq W_+(t)$ for $t \geq a+h$, and in particular $W_+(a) \geq W_+(a+h)$. It follows that $W_+(a+h) = W_+(a)$. Thus the right principal solution is a constant, and the left principal solution is likewise. Applying Theorem 9 we obtain

THEOREM 10. *Suppose* $C \geq 0$ *and condition* [C] *holds for the autonomous equation* (17). *Then* (17) *is disconjugate on the interval* $(-\infty, \infty)$ *if and only if there exists a constant symmetric matrix* W *such that*

$$A + WB + B^*W + WCW = 0 \,. \tag{19}$$

In this case there exist symmetric solutions W_+, W_- *of* (19) *with* $W_+ \leq W_-$ *such that if* $W(t)$ *is any symmetric solution of the Riccati equation* (18) *on the interval* $(-\infty, \infty)$ *then*

$$W_+ \leq W(t) \leq W_- \quad \text{for} \quad -\infty < t < \infty .$$

In particular any symmetric solution W *of* (19) *satisfies* $W_+ \leq W \leq W_-$.

For autonomous equations the condition [C] can be given a purely algebraic form.

LEMMA 9. *The autonomous equation* (17) *satisfies condition* [C] *on some, and thus on every, interval* I *if and only if the* $n \times n^2$ *block matrix*

$$\psi = \left(C \; BC \; \ldots \; B^{n-1}C\right)$$

has rank n .

Proof. Since the equation is autonomous, condition [C] holds on any interval of length h if and only if it holds on the interval $[0, h]$. By the proof of Lemma 3 this will be the case if and only if the matrix

$$R = \int_0^\tau e^{-tB} CC^* e^{-tB^*} dt$$

is positive definite for $0 < \tau \leq h$.

Suppose this is not so. Then there exists a vector $\eta \neq 0$ such that

$$C^* e^{-tB^*} \eta = 0 \quad \text{for} \quad 0 \leq t \leq \tau .$$

Differentiating $n - 1$ times with respect to t and setting $t = 0$ we obtain

$$C^*(B^*)^k \eta = 0 \quad (k = 0, \ldots, n-1) .$$

Thus η is orthogonal to every column of the matrix ψ , which consequently has rank $< n$. Conversely if ψ has rank $< n$ there exists a vector $\eta \neq 0$ satisfying these relations. By the Cayley-Hamilton theorem it follows that $C^* e^{-tB^*} \eta = 0$ for all t . Hence R is not positive definite for any $\tau > 0$.

Finally we will use matrix Riccati equations to obtain an analogue of Theorem 6 of Chapter 1. The proof is based on a lemma of independent interest.

LEMMA 10. *Let* $W(t)$ *be a symmetric solution of the Riccati equation* (14) *on an interval* $[a-\delta, a+\delta]$ *and let* K, α *be positive constants such that throughout this interval*

$$|A(t)| \leq K , \quad |B(t)+B^*(t)| \leq K , \quad C(t) \geq 2\alpha I .$$

Then $|W(a)| \leq M$, *where* M *is a positive constant depending only on* K, α, δ .

Proof. We use the Euclidean norm throughout. If ξ is a unit eigenvector of $W(t)$ belonging to the eigenvalue λ then

$$\begin{aligned}\xi^* W'(t)\xi &= -\ \xi^* A\xi - \lambda\xi^*(B+B^*)\xi - \lambda^2\xi^* C\xi \\ &\le -\ 2\alpha\lambda^2 + K|\lambda| + K \\ &\le -\ \alpha\lambda^2\end{aligned}$$

for $|\lambda| \ge L = 1 + K\alpha^{-1}$.

Let $\mu(t)$ be the least eigenvalue of $W(t)$ and suppose that $\mu(a) < -L$. If in the above argument we take $\lambda = \mu(t)$ then

$$\begin{aligned}\xi^* W(t+h)\xi &= \xi^* W(t)\xi + h[\xi^* W'(t)\xi + o(1)] \\ &\le \mu(t) - h[\alpha\mu^2(t) + o(1)] .\end{aligned}$$

Hence

$$\mu(t+h) \le \mu(t) - h[\alpha\mu^2(t) + o(1)]$$

and

$$D^+\mu \le -\alpha\mu^2 < 0 .$$

Since $\mu(t)$ is a continuous function it follows from the basic result on differential inequalities that for $t > a$, $\mu(t)$ remains less than $-L$ and moreover

$$\mu(t) \le \mu(a)[1 + \alpha\mu(a)(t-a)]^{-1} .$$

Since the right side tends to $-\infty$ as $t \to a - [\alpha\mu(a)]^{-1}$ it follows that $-[\alpha\mu(a)]^{-1} > \delta$, *i.e.*, $-\mu(a) < (\alpha\delta)^{-1}$.

Similarly, by considering values of t less than a , we can show that if $\nu(t)$ is the greatest eigenvalue of $W(t)$ and $\nu(a) > L$ then $\nu(a) < (\alpha\delta)^{-1}$. Hence

$$|W(a)| = \max[-\mu(a), \nu(a)] \le \max[L, (\alpha\delta)^{-1}] .$$

THEOREM 11. *Let $H_\nu(t)$ be a sequence of symmetric matrix functions such that $H_\nu(t) \to H(t)$ uniformly on compact subintervals of the open interval I as $\nu \to \infty$. If each equation*

$$Jx' = H_\nu(t)x \qquad (1)_\nu$$

is disconjugate on I and if the limit equation (1) *satisfies condition* [++] *then it is also disconjugate on I .*

Proof. Let a, b be any two points of I with $a < b$. Let c be the midpoint of $[a, b]$ and choose $\delta' > (b-a)/2$ so that the interval $J = [c-\delta', c+\delta']$ is contained in I . There exists an $\alpha > 0$ such that $C(t) \ge 4\alpha I$ for all t in J . Hence $C_\nu(t) \ge 2\alpha I$ for all t in J if $\nu \ge \nu_0$. We can also choose $K > 0$ so that

$$|A(t)| \le K , \quad |B(t)+B^*(t)| \le K \quad \text{for all} \quad t \quad \text{in} \quad J \quad \text{and all} \quad \nu .$$

By Theorem 1 the Riccati equation $(14)_\nu$ corresponding to $(1)_\nu$ has a symmetric solution $W_\nu(t)$ defined throughout J. By Lemma 9 there exists a positive constant M, depending only on K, α, and $\delta = \delta' - (b-a)/2$ such that

$$|W_\nu(t)| \le M \quad \text{for all} \quad t \quad \text{in} \quad [a, b] \quad \text{and all} \quad \nu \ge \nu_0 .$$

It follows that the derivatives $W'_\nu(t)$ are also uniformly bounded for t in $[a, b]$. Hence, by Ascoli's theorem, the sequence $W_\nu(t)$ contains a subsequence which converges uniformly on $[a, b]$, with limit $W(t)$ say. Then, for the same subsequence,

$$W'_\nu \to - A - WB - B^*W - WCW$$

uniformly on $[a, b]$. It follows that $W(t)$ is differentiable and is a symmetric solution of the Riccati equation (14). Therefore, by Theorem 2, the equation (1) is disconjugate on $[a, b]$.

5. Calculus of variations

We are going to consider a variational problem connected with the linear Hamiltonian equation (1). Suppose the interval $I = [a, b]$ is compact. Let $z(t)$ be any piecewise continuous vector function defined on I and let $y(t)$ be a solution of the inhomogeneous linear equation

$$y' = B(t)y + C(t)z(t) \tag{20}$$

satisfying the boundary conditions

$$y(a) = y(b) = 0 . \tag{21}$$

We will call such pairs y, z *admissible*. The quadratic functional

$$Q = \int_a^b [z^*(t)C(t)z(t)-y^*(t)A(t)y(t)]dt$$

will be said to be *non-negative* if $Q \ge 0$ for all admissible pairs y, z and *positive* if in addition $Q = 0$ implies $y(t) \equiv 0$. The problem is to determine conditions under which Q is non-negative or positive.

If $z(t)$ is continuously differentiable the expression for Q can be transformed. In fact by (20)

$$Q = \int_a^b [(y^*B^*+z^*C)z-y^*(Ay+B^*z)]dt$$

$$= \int_a^b [y'^*z-y^*(Ay+B^*z)]dt$$

and hence by (21)

$$Q = -\int_a^b y^*(z'+Ay+B^*z)dt \; . \tag{22}$$

Using (20) again we get

$$Q = \int_a^b x^*(Jx'-Hx)dt \; . \tag{23}$$

We first establish a sufficient condition for Q to be positive.

THEOREM 12. *If* [+] *and* [R] *hold on the compact interval* $I = [a, b]$ *then* Q *is positive.*

Proof. For any symmetric matrix function $W(t)$ and any solution $y(t)$ of (20) we have

$$z^*Cz - y^*Ay - (y^*Wy)' = (Wy-z)^*C(Wy-z) - y^*(W'+A+WB+B^*W+WCW)y \; . \tag{24}$$

Since [R] holds we can take $W(t)$ to be a solution of the Riccati equation (14). If $y(t)$ satisfies the boundary conditions (21) then by integration we get

$$Q = \int_a^b (Wy-z)^*C(Wy-z)dt \; . \tag{25}$$

Since $C(t) \geq 0$ it follows that $Q \geq 0$, with equality only if $C(Wy-z) \equiv 0$. Then $y(t)$ is a solution of the linear equation

$$y' = [B(t)+C(t)W(t)]y$$

such that $y(a) = 0$, and hence $y(t) \equiv 0$.

We show next that the sufficient condition of Theorem 12 is very close to being necessary.

PROPOSITION 11. *If* Q *is non-negative then* [+] *holds.*

Proof. Suppose on the contrary that $\eta^*C(t)\eta < 0$ for some t in I and some vector η . Then the inequality continues to hold throughout some subinterval $J = [c, d]$. Moreover we can suppose this subinterval so small that the symmetric solution $W(t)$ of (14) which satisfies the initial condition $W(c) = I$ is defined throughout it.

Let $U(t)$ be the fundamental matrix of the linear equation

$$u' = [B(t)+C(t)W(t)]u$$

such that $U(c) = I$ and let $\mu(t)$ be any continuous scalar function for $c \leq t \leq d$ which vanishes at both endpoints. The solution $v(t)$ of the inhomogeneous equation

$$v' = [B(t)+C(t)W(t)]v + \mu(t)C(t)\eta$$

which vanishes at c is given by

$$v(t) = \int_c^t U(t)U^{-1}(s)C(s)\mu(s)\eta ds .$$

It vanishes at d also if and only if

$$\Phi[\mu] \equiv \int_c^d U^{-1}(s)C(s)\mu(s)\eta ds = 0 .$$

Since all functions $\mu(t)$ form an infinite-dimensional vector space and all vectors $\Phi[\mu]$ a finite-dimensional vector space, and since the mapping $\mu \to \Phi[\mu]$ is linear, there exists a nontrivial function $\mu(t)$ such that $\Phi[\mu] = 0$. If we set

$$z(t) = W(t)v(t) + \mu(t)\eta \quad \text{in } J, \quad = 0 \quad \text{elsewhere},$$

$$y(t) = v(t) \quad \text{in } J, \quad = 0 \quad \text{elsewhere},$$

then $y(t)$ is a solution of the differential equation (20) satisfying the boundary conditions (21). However for this choice of y, z, (25) gives

$$\begin{aligned} Q &= \int_c^d (z^*Cz-y^*Ay)dt \\ &= \int_c^d \mu^2(t)\eta^*C(t)\eta dt \\ &< 0 , \end{aligned}$$

which is a contradiction.

It is actually sufficient to assume $Q \geq 0$ for all continuously differentiable functions $z(t)$ since we can take $\mu(t)$ to be a continuously differentiable function which vanishes, together with its derivative, at c and d.

PROPOSITION 12. *If* [C] *holds and* $Q = 0$ *implies* $y(t) \equiv 0$ *then* [D] *holds.*

Proof. Let $y(t)$, $z(t)$ be a solution of (2) with $y(c) = y(d) = 0$, where $a \leq c < d \leq b$. If we redefine $y(t)$ and $z(t)$ to be zero outside $[c, d]$ then $z(t)$ is piecewise continuous and $y(t)$ is a solution of (20) satisfying the boundary conditions (21). Moreover, by (22),

$$Q = \int_c^d (z^*Cz - y^*Ay)dt$$

$$= - \int_c^d y^*(z' + Ay + B^*z)dt$$

$$= 0 \ .$$

Therefore $y(t) = 0$ for $c \le t \le d$. By Lemma 2 this implies that the original solution $y(t)$, $z(t)$ is the trivial one. Thus (1) is disconjugate.

The following theorem does not say anything new, but sums up in a convenient way much of the information we have obtained.

THEOREM 13. *Suppose* $I = [a, b]$ *is a compact interval. If* [+] *and* [D] *hold on* I *then the quadratic functional* Q *is positive. Conversely, if* [C] *holds and* Q *is positive then* [+] *and* [D] *hold.*

Proof. If [+] and [D] hold then [R] holds, by Theorem 1, and Q is positive, by Theorem 12. If [C] holds and Q is positive then [+] holds, by Proposition 11, and [D] holds, by Proposition 12.

For noncompact intervals the controllability condition is not required.

THEOREM 14. *Suppose* I *is an open or half open interval. Then* [+] *and* [D] *hold on* I *if and only if the quadratic functional*

$$Q = \int_a^b (z^*Cz - y^*Ay)dt$$

is non-negative for all a, b *in* I *with* $a < b$ *and all admissible pairs* y, z *on* $[a, b]$.

Proof. If [+] and [D] hold on I then Q is positive for any compact subinterval $[a, b]$, by Theorem 13. Conversely, suppose Q is non-negative for any compact subinterval $[a, b]$. Then [+] holds, by Proposition 11. It remains to show that also [D] holds.

Let y_1, z_1 be a solution of (2) such that $y_1(a) = y_1(c) = 0$, where $a < c$, and suppose, for definiteness, that c is not an endpoint of I. Choose any $b > c$ in I and redefine $y_1(t), z_1(t)$ to be zero for $c < t \le b$. Then y_1, z_1 is an admissible pair for the interval $[a, b]$. Let y_2, z_2 be any admissible pair for this interval. Then

$$y = y_1 + \varepsilon y_2 \ , \quad z = z_1 + \varepsilon z_2$$

is also admissible and the quadratic functional has a corresponding value $Q(\varepsilon) \ge 0$ for all real ε. But

$$Q(0) = \int_a^b (z_1^* C z_1 - y_1^* A y_1)\,dt$$
$$= \int_a^c (z_1^* C z_1 - y_1^* A y_1)\,dt$$
$$= 0 \ ,$$

by (23). Since

$$Q(\varepsilon) - Q(0) = 2\varepsilon \int_a^b (z_2^* C z_1 - y_2^* A y_1)\,dt + O(\varepsilon^2)$$

it follows that

$$\int_a^b (z_2^* C z_1 - y_2^* A y_1)\,dt = 0 \ .$$

Thus

$$[y_2^* z_1]_a^c = \int_a^c (z_2^* C z_1 - y_2^* A y_1)\,dt = 0 \ .$$

Hence $y_2^*(c) z_1(c) = 0$. Since we can choose y_2, z_2 so that $y_2^*(c) = z_1(c)$ this implies that $z_1(c) = 0$. Therefore, since $y_1(c) = 0$, the original solution is the trivial one. This completes the proof.

Just as for scalar equations, one can connect other variational problems with the equation (1). Thus Theorem 8 of Chapter 1 has the following direct extension.

THEOREM 15. *Suppose* $P = P^*$ *and* $C(t) \geq 0$ *on* $I = [a, b]$. *Then the following statements are equivalent:*

(i) the system (2) *has no nontrivial solution* $y(t)$, $z(t)$ *such that*

$$z(a) = Py(z) \ , \quad y(c) = 0 \quad \text{for some} \quad c \in (a, b] \ ,$$

(ii) the Riccati inequality $R[W] \leq 0$ *has a symmetric solution* $W(t)$ *on* I *such that* $W(a) \leq P$,

(iii) the quadratic functional

$$Q = y^*(a) P y(a) + \int_a^b (z^* C z - y^* A y)\,dt$$

is positive for any piecewise continuous vector function $z(t)$ *and any nontrivial solution* $y(t)$ *of the equation*

$$y' = B(t) y + C(t) z(t)$$

such that $y(b) = 0$.

Proof. *(i)* ⇒ *(ii)*. Let Y, Z be the solution of the matrix equation (3) such that $Y(a) = I$, $Z(a) = P$. Then $Y(t)$ is invertible for $a < t \leq b$ and *(ii)* holds for $W(t) = Z(t)Y^{-1}(t)$.

(ii) ⇒ *(iii)*. By (24) we have

$$Q = y^*(a)[P-W(a)]y(a) + \int_a^b (Wy-z)^*C(Wy-z)dt$$
$$\geq 0 ,$$

with equality only if $C(Wy-z) \equiv 0$ and $[P-W(a)]y(a) = 0$. Then $y(t)$ is a solution of the equation

$$y' = (B+CW)y$$

such that $y(b) = 0$. Therefore $y(t) \equiv 0$.

(iii) ⇒ *(i)*. Let $y(t)$, $z(t)$ be a solution of (2) such that $z(a) = Py(a)$, $y(c) = 0$, where $a < c \leq b$. If we redefine $y(t)$, $z(t)$ to be zero for $c < t \leq b$ then

$$\begin{aligned} Q - y^*(a)Py(a) &= \int_a^c (z^*Cz-y^*Ay)dt \\ &= \int_a^c \{(y^*B^*+z^*C)z-y^*(Ay+B^*z)\}dt \\ &= \int_a^c (y'^*z+y^*z')dt \\ &= [y^*z]_a^c \\ &= -y^*(a)Py(a) . \end{aligned}$$

Therefore $Q = 0$ and $y(t) \equiv 0$. In particular $y(a) = 0$ and hence the solution $y(t)$, $z(t)$ is trivial.

Finally we give an extension to systems of the disconjugacy criterion established in Theorem 13 of Chapter 1. In the proof we use the following inequality for positive definite symmetric matrices.

LEMMA 11. *If* $A > 0$, $B > 0$ *then*

$$A^{-1} + B^{-1} \geq 4(A+B)^{-1} . \tag{26}$$

Proof. If $C > 0$ then from $(C+I)^2 \geq 4C$ we obtain

$$C + I \geq 4C(C+I)^{-1} = 4\left(C^{-1}+I\right)^{-1} .$$

With $C = A^{-1}$ this gives (26) with $B = I$. The general case follows on replacing A by $B^{-1/2}AB^{-1/2}$.

PROPOSITION 13. *Let* $P(t)$, $Q(t)$ *be continuous symmetric matrix functions on a compact interval* $I = [a, b]$ *such that* $P(t) > 0$ *and* $Q(t) \le q(t)I$, *where* $q(t)$ *is a continuous non-negative function. If the constant symmetric matrix*

$$D = 4\left[\int_a^b P^{-1}(t)dt\right]^{-1} - \left[\int_a^b q(t)dt\right]I$$

is non-negative then the equation

$$[P(t)y']' + Q(t)y = 0 \tag{27}$$

is disconjugate on I .

Proof. Suppose (27) has a nontrivial solution $u(t)$ such that $u(a_1) = u(b_1) = 0$, where $a \le a_1 < b_1 \le b$. If we put $\eta(t) = u(t)$ for $a_1 \le t \le b_1$, $= 0$ elsewhere in $[a, b]$, then $\eta(t)$ is piecewise continuously differentiable and

$$\begin{aligned} F[\eta] &\equiv \int_a^b (\eta'^*P\eta'-\eta^*Q\eta)dt \\ &= \int_{a_1}^{b_1} (u'^*Pu'-u^*Qu)dt \\ &= 0 . \end{aligned}$$

Choose any c such that $a_1 < c < b_1$ and put

$$Y(t) = \int_a^t P^{-1}(s)ds ,$$

so that $Y'(t) = P^{-1}(t)$. Then $y(t) = Y(t)Y^{-1}(c)\eta(c)$ is a solution of the equation $[P(t)y']' = 0$ such that $y(a) = 0 = \eta(a)$ and $y(c) = \eta(c)$. Hence

$$\int_a^c y'^*P(y'-\eta')dt = 0 ,$$

$$\int_a^c \eta'^*Py'dt = \eta^*(c)Y^{-1}(c)\eta(c) ,$$

and consequently

$$\begin{aligned} 0 &\le \int_a^c (\eta'-y')^*P(\eta'-y')dt \\ &= \int_a^c \eta'^*P\eta'dt - \eta^*(c)Y^{-1}(c)\eta(c) , \end{aligned} \tag{28}$$

with equality only if $P(t)\eta'(t) = Y^{-1}(c)\eta(c)$ for $a \le t \le c$. Similarly, putting

$$Z(t) = -\int_t^b P^{-1}(s)ds$$

we obtain

$$0 \le \int_c^b \eta'^*P\eta'dt + \eta^*(c)Z^{-1}(c)\eta(c) , \tag{29}$$

with equality only if $P(t)\eta'(t) = Z^{-1}(c)\eta(c)$ for $c \le t \le b$. If equality holds in both places then

$$\left[\int_a^c P^{-1}(s)ds\right]^{-1}\eta(c) = -\left[\int_c^b P^{-1}(s)ds\right]^{-1}\eta(c)$$

and hence, since $P^{-1}(s) > 0$, $\eta(c) = 0$. Thus, if $\eta(c) \ne 0$, it follows from (28) and (29) by addition that

$$\int_a^b \eta'^*P\eta'dt > \eta^*(c)[Y^{-1}(c)-Z^{-1}(c)]\eta(c) .$$

Therefore, by Lemma 11 with

$$A = \int_a^c P^{-1}(s)ds , \quad B = \int_c^b P^{-1}(s)ds ,$$

$$\int_a^b \eta'^*P\eta'dt > 4\eta^*(c)\left[\int_a^b P^{-1}(s)ds\right]^{-1}\eta(c) .$$

Also

$$\int_a^b \eta^*Q\eta dt \le \int_a^b q(t)|\eta(t)|^2dt .$$

Take c to be a point at which $|\eta(t)|$ assumes its maximum value on $[a, b]$. Then

$$\int_a^b q(t)|\eta(t)|^2dt \le |\eta(c)|^2 \int_a^b q(t)dt$$

and hence

$$\begin{aligned} F[\eta] &> 4\eta^*(c)\left[\int_a^b P^{-1}(s)ds\right]^{-1}\eta(c) - \left[\int_a^b q(t)dt\right]\eta^*(c)\eta(c) \\ &= \eta^*(c)D\eta(c) \\ &\ge 0 . \end{aligned}$$

Thus we have a contradiction.

The result tells us something new even for scalar equations. If $p(t) > 0$, $q(t) \ge 0$ are continuous functions on $I = [a, b]$ such that

$$\int_a^b dt/p(t) \int_a^b q(t)dt \le 4$$

then the equation

$$[p(t)y']' + q(t)y = 0$$

is disconjugate on I.

We can make the set of all linear Hamiltonian equations (1) on a compact interval I into a metric space by defining the distance between two equations $(1)_1$ and $(1)_2$ to be

$$\sup_{t\in I} |H_1(t)-H_2(t)| .$$

Then we have

PROPOSITION 14. *The set of all disconjugate equations* (1) *satisfying* [++] *on a compact interval* I *is connected and open.*

Proof. If $C > 0$ then (20) can be solved for z and the quadratic functional Q can be written

$$Q = \int_a^b \left\{y'^*C^{-1}y'-y'^*C^{-1}By-y^*B^*C^{-1}y'-y^*\left(A-B^*C^{-1}B\right)y\right\}dt .$$

By Theorem 13 the equation (1) is disconjugate if and only if $Q > 0$ for all nontrivial piecewise continuously differentiable functions $y(t)$ such that $y(a) = y(b) = 0$. Let $(1)_1$ and $(1)_2$ be any two disconjugate equations satisfying [++]. Then the corresponding quadratic functionals Q_1 and Q_2 are positive. Hence $Q = \lambda Q_1 + (1-\lambda)Q_2$ is positive for $0 \le \lambda \le 1$. We can write $[C(\lambda)]^{-1} = \lambda C_1^{-1} + (1-\lambda)C_2^{-1}$, where $C(\lambda) > 0$. If we take

$$B(\lambda) = C(\lambda)\left[\lambda C_1^{-1}B_1+(1-\lambda)C_2^{-1}B_2\right]$$

$$A(\lambda) = B^*(\lambda)C^{-1}(\lambda)B(\lambda) + \lambda\left(A_1-B_1^*C_1^{-1}B_1\right) + (1-\lambda)\left(A_2-B_2^*C_2^{-1}B_2\right) ,$$

then Q is the quadratic functional corresponding to the equation (1). In this way we obtain a path of disconjugate equations satisfying [++] which joins $(1)_1$ to $(1)_2$.

Suppose that for each positive integer ν there is an equation $(1)_\nu$, distant less than $1/\nu$ from (1), which is not disconjugate. Let $x_\nu(t)$ be a solution of $(1)_\nu$ such that $|x_\nu(a)| = 1$ and $y_\nu(a_\nu) = y_\nu(b_\nu) = 0$, where $a \le a_\nu < b_\nu \le b$. By restricting attention to a suitable subsequence we can ensure that $x_\nu(a) \to \xi$ as

$\nu \to \infty$ and $a_\nu \to a_\infty$, $b_\nu \to b_\infty$. Then $x_\nu(t) \to x(t)$ uniformly on I as $\nu \to \infty$, where $x(t)$ is the nontrivial solution of (1) which satisfies the initial condition $x(a) = \xi$. Evidently $y(a_\infty) = y(b_\infty) = 0$. This contradicts the disconjugacy of (1) except possibly if $a_\infty = b_\infty$. Since $x_\nu'(t) \to x'(t)$ uniformly on I, and since each coordinate of $y_\nu'(t)$ vanishes between a_ν and b_ν, in the excepted case $a_\infty = b_\infty$ we have $y'(a_\infty) = 0$. Since $C > 0$ it follows that $z(a_\infty) = 0$. Hence $x(t)$ is the trivial solution and we still have a contradiction. We conclude that there exists $\delta > 0$ such that every equation distant less than δ from (1) is disconjugate. Since $C(t) \geq \alpha I$ throughout I, for some $\alpha > 0$, we can choose δ so that every such equation also satisfies [++].

6. The method of polar coordinates

We have seen that the use of polar coordinates enabled us to establish the most general form of the comparison theorem for scalar equations. This method can also be extended to systems, although the extension which we will describe is not a direct generalisation of the method used in Chapter 1. Since it requires no extra effort we will also allow more general boundary conditions than those which appear in the definition of disconjugacy.

The proof of the comparison theorem will be based on a property of the eigenvalues of unitary matrix functions. This property will first be deduced from an analogous result for Hermitian matrix functions.

LEMMA 12. *Let $A(t)$ be a Hermitian matrix function which is differentiable at $t = c$. If $\xi^*A'(c)\xi > 0$ for all eigenvectors ξ of $A(c)$ belonging to the eigenvalue λ then the eigenvalues of $A(t)$ in the neighbourhood of λ for values of t slightly greater (less) than c are greater (less) than λ.*

Proof. We consider the case $t > c$ only, the changes required for $t < c$ being obvious. Suppose that λ is an eigenvalue of $A(c)$ of multiplicity r and that, taking account of multiplicity, there exist p eigenvalues greater than λ. There exists $\mu > 0$ such that

$$\xi^*A'(c)\xi \geq \mu\xi^*\xi$$

for all eigenvectors ξ of $A(c)$ belonging to the eigenvalue λ. Then for all such ξ

$$\begin{aligned}\xi^*A(c+h)\xi &= \xi^*A(c)\xi + h\xi^*[A'(c)+o(1)]\xi \\ &\geq (\lambda+\tfrac{1}{2}\mu h)\xi^*\xi ,\end{aligned}$$

if h is positive and sufficiently small. The same inequality certainly holds for all eigenvectors corresponding to eigenvalues of $A(c)$ greater than λ . Hence it holds on a subspace of dimension $p+r$. Any subspace of dimension $n-(p+r)+1$ has non-empty intersection with this. Therefore, by the minimax principle, $A(c+h)$ has $p+r$ eigenvalues greater than $\lambda+\frac{1}{2}\mu h$ for sufficiently small $h>0$. Since the eigenvalues of $A(t)$ are continuous at $t=c$ this completes the proof.

LEMMA 13. *Let the unitary matrix* $\Omega(t)$ *satisfy the differential equation*

$$\Omega' = i\Omega R(t) ,$$

where $R(t)$ *is a continuous Hermitian matrix for* $a \le t \le b$. *If* $a \le c \le b$ *and* $\xi^* R(c)\xi > 0$ *for all eigenvectors* ξ *of* $\Omega(c)$ *belonging to the eigenvalue* $\omega = e^{i\gamma}$ *then the eigenvalues of* $\Omega(t)$ *in the neighbourhood of* ω *for values of* t *slightly greater (less) than* c *have argument greater (less) than* γ .

Proof. Choose α real so that $e^{i\alpha}$ is not an eigenvalue of $\Omega(c)$. Then

$$A(t) = i\left[e^{i\alpha}I+\Omega(t)\right]\left[e^{i\alpha}I-\Omega(t)\right]^{-1}$$

is defined in the neighbourhood of c . Since Ω is unitary

$$\left(e^{i\alpha}I-\Omega\right)^{*-1} = -\,e^{i\alpha}\Omega\left(e^{i\alpha}I-\Omega\right)^{-1}$$

and hence A is Hermitian. Since

$$A = 2ie^{i\alpha}\left(e^{i\alpha}I-\Omega\right)^{-1} - iI ,$$

we have

$$\begin{aligned} A' &= -\,2e^{i\alpha}\left(e^{i\alpha}I-\Omega\right)^{-1}\Omega R\left(e^{i\alpha}I-\Omega\right)^{-1} \\ &= 2\left(e^{i\alpha}I-\Omega\right)^{*-1}R\left(e^{i\alpha}I-\Omega\right)^{-1} . \end{aligned}$$

The eigenvalues $\tilde{\lambda}$ of A are connected with the eigenvalues $\tilde{\omega} = e^{i\beta}$ of Ω by the relation

$$\tilde{\lambda} = i\left(e^{i\alpha}+\tilde{\omega}\right)\left(e^{i\alpha}-\tilde{\omega}\right)^{-1} = \cot(\alpha-\beta)/2 .$$

Moreover if ξ is an eigenvector of Ω belonging to the eigenvalue $\tilde{\omega}$ then it is also an eigenvector of A belonging to the eigenvalue $\tilde{\lambda}$. Hence $\xi^* A'(c)\xi > 0$ for all eigenvectors ξ of $A(c)$ belonging to the eigenvalue $\lambda = \cot(\alpha-\gamma)/2$. Thus we have the situation of Lemma 12. Since β increases at the same time as $\tilde{\lambda}$ the result follows.

Consider now the linear Hamiltonian equation

$$Jx' = H(t)x \tag{1}$$

on the interval $I=[a, b]$. Let $X(t)$ be the fundamental matrix of (1) such that $X(a) = I$. Then $X(t)$ is a symplectic matrix, *i.e.*, it satisfies

$$X^*JX = J \tag{30}$$

for each t in I . In fact if we differentiate the left side and take account of (1) we find that its derivative is 0 .

Let M, N be constant matrices such that

$$M^*JM = N^*JN \tag{31}$$

and

$$M\xi = N\xi = 0 \quad \text{implies} \quad \xi = 0 . \tag{32}$$

We look for solutions $x(t)$ of (1) satisfying the *self-adjoint* boundary conditions

$$x(a) = M\xi , \quad x(c) = N\xi \tag{33}$$

for some vector ξ and some point c in I . If the boundary value problem (1) - (33) has a nontrivial solution $x(t)$ we will say that c is a *conjugate point* of a relative to the given boundary conditions. The number of linearly independent solutions is called the *order* of the conjugate point. If c is not a conjugate point we ascribe to it the order zero.

For example, we can take

$$M = \begin{pmatrix} 0 & 0 \\ I_n & 0 \end{pmatrix} , \quad N = \begin{pmatrix} 0 & 0 \\ 0 & I_n \end{pmatrix} .$$

In this case conjugate points with respect to the boundary conditions (33) are the same as conjugate points in the sense of Section 3.

Put

$$U = J(XM-N) , \quad V = XM + N , \tag{34}$$

$$\Omega = (V+iU)(V-iU)^{-1} . \tag{35}$$

PROPOSITION 15. *The matrix* $\Omega(t)$ *exists and is unitary for all* t *in* I . *It satisfies the differential equation*

$$\Omega' = i\Omega R(t) , \tag{36}$$

where the Hermitian matrix $R(t)$ *is given by*

$$R = 4(V^*+iU^*)^{-1}M^*X^*HXM(V-iU)^{-1} .$$

Proof. We show first that

$$V^*U - U^*V = 0 .$$

In fact, since $J^* = -J$, the left side is equal to

$$\begin{aligned}(M^*X^*+N^*)J(XM-N) + (M^*X^*-N^*)J(XM+N) &= 2(M^*X^*JXM-N^*JN)\\ &= 0 ,\end{aligned}$$

by (30) and (31). Hence

$$(V^*-iU^*)(V+iU) = (V^*+iU^*)(V-iU) = V^*V + U^*U \ . \tag{37}$$

The right side is a non-negative symmetric matrix. If for some t in I and some vector ξ

$$\xi^*(V^*V+U^*U)\xi = 0$$

then $U\xi = V\xi = 0$ and hence $XM\xi = N\xi = 0$. Since X is invertible this implies $\xi = 0$ by (32). Thus

$$V^*V + U^*U > 0 \ .$$

It follows from (37) that $V - iU$ is invertible, *i.e.*, Ω exists, and

$$\Omega^*\Omega = (V^*+iU^*)^{-1}(V^*-iU^*)(V+iU)(V-iU)^{-1} = I \ .$$

Finally

$$\begin{aligned}\Omega' &= (V'+iU')(V-iU)^{-1} - (V+iU)(V-iU)^{-1}(V'-iU')(V-iU)^{-1} \\ &= [(V'+iU')-\Omega(V'-iU')](V-iU)^{-1} \\ &= \Omega(V^*+iU^*)^{-1}[(V^*-iU^*)(V'+iU')-(V^*+iU^*)(V'-iU')](V-iU)^{-1} \ ,\end{aligned}$$

because Ω is unitary. Thus

$$\Omega' = 2i\Omega(V^*+iU^*)^{-1}(V^*U'-U^*V')(V-iU)^{-1} \ . \tag{38}$$

But by (34) and (1)

$$\begin{aligned}V^*U' - U^*V' &= (M^*X^*+N^*)JX'M + (M^*X^*-N^*)JX'M \\ &= 2M^*X^*HXM \ .\end{aligned}$$

PROPOSITION 16. *The order of* c $(a \le c \le b)$ *is equal to the number of eigenvalues of* $\Omega(c)$ *equal to* $+1$.

Proof. We show first that $U\xi = 0$ if and only if $\zeta = (V-iU)\xi$ satisfies $\Omega\zeta = \zeta$. In fact if $U\xi = 0$ then

$$\Omega\zeta = (V+iU)\xi = V\xi = \zeta \ .$$

Conversely, if $\Omega\zeta = \zeta$ then

$$(V+iU)\xi = \Omega\zeta = \zeta = (V-iU)\xi$$

and hence $U\xi = 0$.

Now the solution of (1) which satisfies the first boundary condition (33) is $x(t) = X(t)M\xi$. This satisfies the second boundary condition also if and only if $[X(c)M-N]\xi = 0$, *i.e.*, $U(c)\xi = 0$. Moreover the solution $x(t)$ is identically zero if and only if $M\xi = N\xi = 0$, *i.e.*, $\xi = 0$. Hence the order of c is equal to the dimension of the nullspace of $U(c)$ and therefore, since $V - iU$ is invertible, to the dimension of the eigenspace of $\Omega(c)$ corresponding to the eigenvalue 1 .

For the remainder of this section we assume that the coefficient matrix $H(t)$ satisfies the following condition:

[P] $N\xi \neq 0$ implies $\xi^*N^*H(t)N\xi > 0$.

With its aid we can prove

PROPOSITION 17. *The interval* $I = [a, b]$ *contains at most finitely many conjugate points of* a .

Proof. We show first that the set of conjugate points is closed. Let $\{t_\nu\}$ be a sequence of conjugate points such that $t_\nu \to c$ as $\nu \to \infty$, where $a \leq c \leq b$. Then the equation (1) has a solution $x_\nu(t) = X(t)M\xi_\nu$, where $\xi_\nu^*\xi_\nu = 1$, such that $x_\nu(t_\nu) = N\xi_\nu$. By restricting consideration to a suitable subsequence we may suppose that $\xi_\nu \to \tilde{\xi}$, where $\tilde{\xi}^*\tilde{\xi} = 1$. Then $x_\nu(t)$ converges uniformly to $\tilde{x}(t) = X(t)M\tilde{\xi}$, and $\tilde{x}(c) = N\tilde{\xi}$. Hence c is also a conjugate point.

We show next that the conjugate points are isolated. Suppose c is a conjugate point. For any eigenvector ζ of $\Omega(c)$ belonging to the eigenvalue 1 we have $\zeta^*R(c)\zeta > 0$. For if $\xi = (V-iU)^{-1}\zeta$ then $U\xi = 0$, *i.e.*, $XM\xi = N\xi$. Hence $N\xi \neq 0$ and, since [P] holds,

$$\zeta^*R\zeta = 4\xi^*M^*X^*HXM\xi = 4\xi^*N^*HN\xi > 0 .$$

Applying Lemma 13 we see that $\Omega(t)$ does not have the eigenvalue 1 for values of t near c and distinct from c . Therefore, by Proposition 16, there are no other conjugate points near c . Since the set of conjugate points is closed and has no limit points it must be finite.

For any c , $a < c \leq b$, we define the *index* $j(c)$ of c to be the sum of the orders of all conjugate points of a in the interval $(a, c]$. Our comparison theorem relates the indices of two different Hamiltonian systems.

THEOREM 16. *Let* $H_1(t)$, $H_2(t)$ *be continuous symmetric matrix functions satisfying the condition* [P] *such that* $H_1(t) \leq H_2(t)$ *for* $a \leq t \leq b$. *Then for any* c , $(a < c \leq b)$, *we have*

$$j_1(c) \leq j_2(c) . \tag{39}$$

Moreover $j_1(c) \leq j_2(c-0)$ *if* $x(t) \equiv 0$ *is the only solution of the equation* $(1)_1$ *satisfying the boundary conditions* (33) *such that* $H_1(t)x(t) = H_2(t)x(t)$ *for* $a \leq t \leq c$.

Proof. Put $H_0 = H_2 - H_1$, and consider the Hamiltonian equation

$$Jx' = (\lambda H_0 + H_1)x . \tag{40}$$

The functions X and Ω are now differentiable functions of the parameter λ, and (38) still holds if derivatives with respect to t are replaced by derivatives with respect to λ, *i.e.*,

$$\Omega_\lambda = 2i\Omega(V^*+iU^*)^{-1}\left(V^*U_\lambda-U^*V_\lambda\right)(V-iU)^{-1} . \tag{41}$$

But

$$\begin{aligned} V^*U_\lambda - U^*V_\lambda &= (M^*X^*+N^*)JX_\lambda M + (M^*X^*-N^*)JX_\lambda M \\ &= 2M^*X^*JX_\lambda M . \end{aligned} \tag{42}$$

Differentiating (40) we see that X_λ is a solution of the inhomogeneous equation

$$JX_\lambda' = \left(\lambda H_0+H_1\right)X_\lambda + H_0X$$

satisfying the initial condition $X_\lambda(a) = 0$. Therefore, by the variation of constants formula and (30),

$$\begin{aligned} X_\lambda(t) &= \int_a^t X(t)X^{-1}(s)J^{-1}H_0(s)X(s)ds \\ &= X(t)J^{-1} \int_a^t X^*(s)H_0(s)X(s)ds . \end{aligned}$$

Hence

$$X^*(t)JX_\lambda(t) = \int_a^t X^*(s)H_0(s)X(s)ds .$$

Combining this with (41) and (42) we see that Ω is a solution of the differential equation

$$\Omega_\lambda = i\Omega Q ,$$

where the Hermitian matrix $Q = Q(t, \lambda)$ is given by

$$Q = 4(V^*+iU^*)^{-1}M^*\left[\int_a^t X^*(s)H_0(s)X(s)ds\right]M(V-iU)^{-1} .$$

Since $H_0 \geq 0$ we have $Q \geq 0$. It follows by a limiting process from Lemma 13 that for each eigenvalue $\omega_k(t, \lambda)$ of $\Omega(t, \lambda)$ the argument of $\omega_k(t, \lambda)$ is a nondecreasing function of λ. Thus

$$\arg \omega_k(t, 1) \geq \arg \omega_k(t, 0) .$$

But both sides agree for $t = a$ and, by the proof of Proposition 17, each increases with t at a multiple of 2π. Hence the number of times $n_k(1)$ that $\arg \omega_k(t, 1)$ is equal to a multiple of 2π on the interval $(a, c]$ is at least as great as the number of times $n_k(0)$ that $\arg\omega_k(t, 0)$ is equal to a multiple of 2π on this interval. Therefore

$$j_2(c) = \sum_k n_k(1) \geq \sum_k n_k(0) = j_1(c) .$$

We will have $j_2(c-0) \geq j_1(c)$ if $\arg\omega_k(c, 1) > \arg\omega_k(c, 0)$ for every k for which $\arg\omega_k(c, 0)$ is a multiple of 2π. This inequality in turn will follow from Lemma 13 if $\zeta^*Q(c, 0)\zeta > 0$ for all eigenvalues ζ of $\Omega(c, 0)$ belonging to the eigenvalue 1. But $\zeta^*Q\zeta = 0$ implies

$$H_0(s)X(s)M(V-iU)^{-1}\zeta = 0$$

for $a \leq s \leq c$. Thus, putting $\xi = (V-iU)^{-1}\zeta$, we see that $x(t) = X(t)M\xi$ is a solution of the differential equation (39) such that $H_0(t)x(t) \equiv 0$. Moreover, $U\xi = 0$, since $\Omega\zeta = \zeta$. Hence $x(c) = N\xi$ and $x(t)$ satisfies the boundary conditions (33). Therefore, by the hypothesis of the theorem, $x(t) \equiv 0$. Thus $M\xi = N\xi = 0$, which implies $\xi = 0$ and $\zeta = 0$. But this contradicts the assumption that ζ is an eigenvector.

If

$$N = \begin{pmatrix} 0 & 0 \\ 0 & I_n \end{pmatrix}$$

then the condition [P] is equivalent to the condition [++]. Thus in particular, we have shown that if $H_2(t) \geq H_1(t)$ and $C_1(t) > 0$ for all t in $I = [a, b]$ then the equation $(1)_2$ has at least as many points in I conjugate to a as has the equation $(1)_1$, where conjugate points are understood in the sense of Section 3 but are counted according to their order.

7. Self-adjoint equations of higher order

Let

$$L(u) \equiv p_n(t)u^{(n)} + p_{n-1}(t)u^{(n-1)} + \dots + p_0(t)u \tag{43}$$

be a linear differential operator of order n with continuous coefficients $p_k(t)$ and leading coefficient $p_n(t) \neq 0$. We say that another such operator

$$L^*(v) \equiv q_n(t)v^{(n)} + q_{n-1}(t)v^{(n-1)} + \dots + q_0(t)v$$

is its *adjoint* if for all $u, v \in C^n$

$$vL(u) - uL^*(v) = \frac{d}{dt} B(u, v) , \tag{44}$$

where

$$B(u, v) = \sum_{0 \le j+k<n} b_{jk}(t) u^{(j)} v^{(k)}$$

is a bilinear form with continuously differentiable coefficients $b_{jk}(t)$.

We show first that for given L the relation (44) uniquely determines both L^* and B . It is sufficient to show that

$$uL^*(v) = \frac{d}{dt} B(u, v) \tag{45}$$

for all $u, v \in C^n$ implies $L^* \equiv 0$, $B \equiv 0$. We can write B in the form,

$$B = \sum_{j=0}^{n-1} c_j u^{(j)} ,$$

where

$$c_j = \sum_{k=0}^{n-1-j} b_{jk} v^{(k)} .$$

Then

$$dB/dt = c_{n-1} u^{(n)} + (c_{n-2} + c'_{n-1}) u^{(n-1)} + \dots + (c_0 + c'_1) u' + c'_0 u .$$

Thus for (45) to hold for all $u \in C^n$ we must have $c_j = 0$ $(j = 0, \dots, n-1)$. But $c_j = 0$ for all $v \in C^n$ implies $b_{jk} = 0$ $(k = 0, \dots, n-1-j)$. Hence $L^* \equiv 0$ and $B \equiv 0$.

We show next that any linear differential operator L with coefficients $p_k(t) \in C^k$ has an adjoint. If $u, v \in C^k$ then

$$\frac{d}{dt}\left\{\sum_{j=0}^{k-1} (-1)^j u^{(k-1-j)} v^{(j)}\right\} = u^{(k)} v + (-1)^{k-1} u v^{(k)} , \tag{46}$$

since on differentiating by the product rule the remaining terms cancel. Replacing v by $p_k(t)v$ and summing from $k = 0, \dots, n$ we obtain a relation of the form (44), where

$$L^*(v) = (-1)^n [p_n(t)v]^{(n)} + (-1)^{n-1} [p_{n-1}(t)v]^{(n-1)} + \dots + p_0 v \tag{47}$$

and

$$B(u, v) = \sum_{k=0}^{n} \sum_{j=0}^{k-1} (-1)^j u^{(k-j-1)} (p_k v)^{(j)} \; . \tag{48}$$

The adjoint operation is evidently linear:

$$(L_1 + L_2)^* = L_1^* + L_2^* \; ,$$

$$(\lambda L)^* = \lambda L^*$$

for any real number λ. We say that L is *self-adjoint* if $L = L^*$. For example, we will show that the operator

$$L(u) = [pu^{(k)}]^{(k)}$$

of order $2k$ is self-adjoint. If in (46) we replace u by $pu^{(k)}$ we obtain

$$v[pu^{(k)}]^{(k)} = (-1)^k pu^{(k)} v^{(k)} + \frac{d}{dt}\left\{\sum_{j=0}^{k-1} (-1)^j v^{(j)} [pu^{(k)}]^{(k-1-j)}\right\} .$$

Interchanging u and v and subtracting we get

$$v[pu^{(k)}]^{(k)} - u[pv^{(k)}]^{(k)} = \frac{d}{dt} \sum_{j=0}^{k-1} (-1)^j \left\{v^{(j)} [pu^{(k)}]^{(k-1-j)} - u^{(j)} [pv^{(k)}]^{(k-1-j)}\right\} .$$

It follows that any operator of the form

$$\tilde{L}(u) = \sum_{k=0}^{n} (-1)^k \left[p_k u^{(k)}\right]^{(k)} \tag{49}$$

is self-adjoint. In this case

$$v\tilde{L}(u) - u\tilde{L}(v) = \frac{d}{dt} \tilde{B}(u, v) \; , \tag{50}$$

where

$$\tilde{B}(u, v) = \sum_{k=0}^{n} \sum_{j=0}^{k-1} (-1)^{k-j} \left\{v^{(j)} \left[p_k u^{(k)}\right]^{(k-1-j)} - u^{(j)} \left[p_k v^{(k)}\right]^{(k-1-j)}\right\} .$$

If we set

$$\begin{aligned} z_k &= \sum_{j=k}^{n} (-1)^{j-k} \left[p_j u^{(j)}\right]^{(j-k)} \\ w_k &= \sum_{j=k}^{n} (-1)^{j-k} \left[p_j v^{(j)}\right]^{(j-k)} \; , \end{aligned} \tag{51}$$

we can write this in the form

$$\tilde{B}(u,\, v) = \sum_{k=1}^{n} \left\{ w_k u^{(k-1)} - z_k v^{(k-1)} \right\} . \tag{52}$$

We show finally that any self-adjoint operator can be expressed in the form (49). If L, given by (43), is to coincide with L^*, given by (47), we must have $(-1)^n = 1$ and, by Leibnitz' product formula,

$$np'_n - p_{n-1} = p_{n-1} .$$

Thus n must be even, $n = 2m$ say, and $p_{n-1} = mp'_n$. Therefore $L(u) - \left[p_n u^{(m)}\right]^{(m)}$ is a self-adjoint operator of order $n - 2$ at most, and the result follows by induction on n.

We will now consider the self-adjoint linear differential equation of order $2n$

$$\tilde{L}(u) \equiv \sum_{k=0}^{n} (-1)^k \left[p_k u^{(k)}\right]^{(k)} = 0 . \tag{53}$$

We assume that $p_k(t) \in C^k$ $(k = 0, 1, \ldots, n)$ and $p_n(t) > 0$ for all t in an interval I. It is not difficult to verify that the equation (53) is equivalent to a Hamiltonian system (2), where

$$y = \begin{pmatrix} u \\ u' \\ \cdot \\ \cdot \\ \cdot \\ u^{(n-1)} \end{pmatrix}, \quad z = \begin{pmatrix} z_1 \\ z_2 \\ \cdot \\ \cdot \\ \cdot \\ z_n \end{pmatrix},$$

with z_k given by (51) and

$$A = -\operatorname{diag}[p_0, p_1, \ldots, p_{n-1}] ,$$

$$C = p_n^{-1} \operatorname{diag}[0, \ldots, 0, 1] ,$$

$$B = \begin{pmatrix} 0 & 1 & 0 & \ldots & 0 \\ 0 & 0 & 1 & \ldots & 0 \\ & & \ldots\ldots & & \\ 0 & 0 & 0 & \ldots & 1 \\ 0 & 0 & 0 & \ldots & 0 \end{pmatrix} .$$

Condition [+] holds for this system because $p_n(t)$ is positive. Condition [C] also holds, since for any solution $y(t)$, $z(t)$ of (2) $y = 0$ throughout a subinterval J implies $z = 0$ throughout J. Our definition of disconjugacy for the system (2) reduces in the present case to the following:

Two points a, b *of* I *are conjugate with respect to the equation* (53) *if there exists a nontrivial solution of* (53) *which has* n*-fold zeros at* a *and* b . *The equation* (53) *is disconjugate on* I *if no two points of* I *are conjugate.*

It follows from Lemma 1, or from (50) and (52), that for any two solutions u, v of (53)

$$\{u, v\} \equiv \sum_{k=1}^{n} \left\{w_k u^{(k-1)} - z_k v^{(k-1)}\right\} = \text{const.} , \tag{54}$$

where z_k and w_k are defined by (51). The solutions u, v will be said to be *conjugate* if this constant is 0 . Evidently any solution is conjugate to itself.

If Y, Z is a solution of the corresponding matrix system (3) then Y is the Wronskian matrix of n solutions $u_1, \ldots, u_n$ of (53). The solution Y, Z is isotropic if and only if u_j is conjugate to u_k $(j, k = 1, \ldots, n)$. By Theorems 1 and 2 we have

THEOREM 17. *The equation* (53) *is disconjugate on a compact interval* I *if and only if there exist* n *mutually conjugate solutions whose Wronskian never vanishes on* I .

It is easily verified that with the present values of A, B, C the quadratic functional Q is given by

$$Q[u] = \int_a^b \sum_{k=0}^{n} p_k (u^{(k)})^2 dt . \tag{55}$$

A function u is admissible if its $(2n-1)$-st derivative exists and is piecewise continuous and if it has n-fold zeros at a and b . The alternative representation (22) shows that if $u \in C^{2n}$ then

$$Q[u] = \int_a^b u\tilde{L}(u) dt . \tag{56}$$

By Theorem 13 we have

THEOREM 18. *The equation* (53) *is disconjugate on a compact interval* $I = [a, b]$ *if and only if the quadratic functional* Q *is positive on the class of admissible functions.*

Suppose now that we have two equations

$$\tilde{L}_1(u) \equiv \sum_{k=0}^{n} (-1)^k \left[p_{k,1} u^{(k)}\right]^{(k)} = 0 , \tag{53$_1$}$$

$$\tilde{L}_2(u) \equiv \sum_{k=0}^{n} (-1)^k \left[p_{k,2} u^{(k)}\right]^{(k)} = 0 . \tag{53$_2$}$$

From Theorem 18 we obtain at once the

COROLLARIES.

(i) If both equations $(53)_1$ *and* $(53)_2$ *are disconjugate on an interval* I *and if*

$$p_k(t) = \lambda_1 p_{k,1}(t) + \lambda_2 p_{k,2}(t) \quad (k = 0, \ldots, n) ,$$

where $\lambda_1 > 0$, $\lambda_2 > 0$, *then the equation* (53) *is also disconjugate on* I .

(ii) If the equation $(53)_2$ *is disconjugate on an interval* I *and*

$$p_{k,1}(t) \geq p_{k,2}(t) \quad (k = 0, \ldots, n) ,$$

then the equation $(53)_1$ *is also disconjugate on* I .

We consider next the factorisation of self-adjoint operators of higher order. Suppose the equation (53) is disconjugate on the compact interval $I = [a, b]$, and hence the quadratic functional $Q[u]$ is positive on the class of admissible functions. Under these circumstances it follows from Lemma 8 that

$$Q[u] = \int_a^b p_n^{-1}\left\{p_n u^{(n)} - w_{nn} u^{(n-1)} - \ldots - w_{n1} u\right\}^2 dt ,$$

where $w_{n1}, \ldots, w_{nn}$ is the last row of the matrix W . That is,

$$Q[u] = \int_a^b [L_1(u)]^2 dt , \tag{57}$$

where

$$L_1(u) = q_n(t)u^{(n)} + q_{n-1}(t)u^{(n-1)} + \ldots + q_0(t)u$$

is a linear differential operator of order n with leading coefficient $q_n = p_n^{1/2}$.

From the expression (50) it follows that

$$\int_a^b \{v\tilde{L}(u) - u\tilde{L}(v)\}dt = 0$$

for any two functions $u, v \in C^{2n}$ with n-fold zeros at a and b . Therefore, by (56),

$$Q[u+v] = Q[u] + Q[v] + 2\int_a^b v\tilde{L}(u)dt .$$

Polarising (57) in the same way and subtracting we get

$$\int_a^b v\tilde{L}(u)dt = \int_a^b L_1(u)L_1(v)dt \; .$$

On the other hand, if v has n-fold zeros at a and b then by (44)

$$\int_a^b \{wL_1(v)-vL_1^*(w)\}dt = 0 \; .$$

Taking $w = L_1(u)$ we deduce that

$$\int_a^b v\tilde{L}(u)dt = \int_a^b vL_1^*[L_1(u)]dt \; .$$

Since this holds for all $v \in C^{2n}$ satisfying the boundary conditions it follows that

$$\tilde{L}(u) = L_1^*[L_1(u)]$$

for all $u \in C^{2n}$ satisfying the boundary conditions. Hence $\tilde{L} \equiv L_1^*L_1$, *i.e.*, when expanded the differential operator on the right side has the same coefficients as the differential operator on the left side.

Conversely, if $\tilde{L} \equiv L_1^*L_1$ and if u is a nontrivial solution of (53) with n-fold zeros at c and d $(c < d)$, then

$$\int_c^d [L_1(u)]^2dt = \int_c^d u\tilde{L}(u)dt = 0 \; .$$

Hence $L_1(u) \equiv 0$ and $u \equiv 0$. Thus the equation (53) is disconjugate on I .

The derivation of (57) from (56) shows that there is an operator L_1 for any symmetric solution W of the corresponding Riccati equation (14), *i.e.*, for any isotropic solution Y, Z of the corresponding system (3) for which Y is invertible, *i.e.*, for any pairwise conjugate solutions $u_1, \ldots, u_n$ of (53) whose Wronskian

$$W(u_1, \ldots, u_n) = \begin{vmatrix} u_1 & u_2 & \cdots & u_n \\ u_1' & u_2' & \cdots & u_n' \\ & \cdots\cdots & & \\ u_1^{(n-1)} & u_2^{(n-1)} & \cdots & u_n^{(n-1)} \end{vmatrix}$$

does not vanish. Moreover $u_1, \ldots, u_n$ will be a fundamental system of solutions of the equation $L_1(u) = 0$. But the differential equation with $u_1, \ldots, u_n$ as fundamental system of solutions and with leading coefficient $p_n^{1/2}$ is uniquely determined and can be explicitly represented in the form

$$L_1(u) = \frac{p_n^{1/2}}{W(u_1,\ldots,u_n)} \begin{vmatrix} u_1 & \ldots & u_n & u \\ u_1' & \ldots & u_n' & u' \\ & \ldots\ldots & & \\ u_1^{(n)} & \ldots & u_n^{(n)} & u^{(n)} \end{vmatrix} . \tag{58}$$

Thus we obtain

THEOREM 19. *The equation* (53) *is disconjugate on a compact interval* I *if and only if the self-adjoint linear differential operator* $\tilde{L}$ *has a factorisation* $\tilde{L} = L_1^* L_1$ *, where* L_1 *is a linear differential operator of order* n *and* L_1^* *its adjoint. Such a factorisation holds if and only if* L_1 *has the form* (58), *where* $u_1, \ldots, u_n$ *are pairwise conjugate solutions of* (53) *whose Wronskian* $W(u_1, \ldots, u_n)$ *does not vanish on* I .

By Theorem 14 the equation (53) is disconjugate on a noncompact interval I if and only if the quadratic functional $Q[u]$ is nonnegative for all a, b in I and all admissible functions u on $[a, b]$. It follows that if a sequence of equations $(53)_\nu$, each disconjugate on a noncompact interval I , is such that $p_{k,\nu}(t) \to p_k(t)$ uniformly on compact subintervals of I as $\nu \to \infty$ then the limit equation (53) is also disconjugate on I . This is the analogue of Theorem 11, which is not applicable here because the condition [++] is not satisfied.

CHAPTER 3

EQUATIONS OF ARBITRARY ORDER

1. General properties of zeros

Let

$$L[y] \equiv y^{(n)} + p_1(t)y^{(n-1)} + \ldots + p_n(t)y = 0 \tag{1}$$

be an n-th order linear differential equation whose coefficients $p_k(t)$ $(k = 1, \ldots, n)$ are continuous on an interval I.

PROPOSITION 1. *If* $I = [a, b]$ *is a compact interval there exists a* $\delta > 0$ *such that* (1) *is disconjugate on every subinterval of length* $< \delta$.

Proof. Put

$$M = \max_{1 \le i \le n} \max_{t \in I} |p_i(t)|$$

and take $\delta = \min[1, 1/nM]$. A sharper estimate for δ is provided by Theorem 1 below. If a solution $y(t)$ has n zeros on a subinterval J of length $< \delta$ then $y^{(k)}(t)$ has at least $n - k$ zeros on J $(k = 1, \ldots, n-1)$. If we put

$$\mu_k = \max_{t \in J} |y^{(k)}(t)|$$

then, by the mean value theorem, $\mu_k \le \delta\mu_{k+1}$ with strict inequality if $\mu_k > 0$. Since $\mu_0 > 0$ it follows that $0 < \mu_k < \delta^{n-k}\mu_n$ $(k = 0, \ldots, n-1)$. On the other hand, from (1)

$$\begin{aligned} \mu_n &\le M(\mu_{n-1} + \ldots + \mu_0) \\ &< M(\delta + \ldots + \delta^n)\mu_n \\ &\le nM\delta\mu_n . \end{aligned}$$

Therefore $1 < nM\delta$, which is a contradiction.

It follows that, if I is compact, there exists a positive integer N such that every nontrivial solution of (1) has less than N zeros on I.

PROPOSITION 2. *For any interval* I *there exists a nontrivial solution of* (1) *with* $n - 1$ *arbitrarily prescribed zeros.*

Proof. Let $y_1(t), \ldots, y_n(t)$ be a fundamental system of solutions of (1).

The solution $y(t) = \alpha_1 y_1(t) + \ldots + \alpha_n y_n(t)$ will have $n - 1$ prescribed zeros if the coefficients $\alpha_1, \ldots, \alpha_n$ satisfy a system of $n - 1$ homogeneous linear equations. Since $n-1 < n$ these equations have a nontrivial solution.

PROPOSITION 3. *If the interval* $I = (a, b)$ *is open and every nontrivial solution has less than* n *distinct zeros then every such solution has less than* n *zeros, counting multiplicities.*

Proof. Suppose on the contrary that there exists a solution with zeros of total multiplicity $\geq n$ and among all such solutions let $y(t)$ be one with the greatest number of distinct zeros. Suppose $y(t)$ has h $(1 \leq h < n)$ distinct zeros t_k of multiplicities r_k $(1 \leq r_k < n)$, where $a < t_1 < \ldots < t_h < b$ and $r = r_1 + \ldots + r_h \geq n$. Choose t_0, t_{h+1} so that $a < t_0 < t_1$ and $t_h < t_{h+1} < b$.

We consider two cases:

(i) $h = n - 1$.

Suppose first that r_k is even for some k $(1 \leq k \leq h)$. Without loss of generality we can assume $y^{(r_k)}(t_k) > 0$. There exists a unique solution $u(t)$ such that $u(t_k) = 1$ and $u(t_i) = 0$ $(i = 1, \ldots, h+1;\ i \neq k)$. The nontrivial solution $y_\varepsilon(t) = y(t) - \varepsilon u(t)$ has $n - 2$ zeros at t_i $(i = 1, \ldots, h;\ i \neq k)$. If we choose α, β so that $t_{k-1} < \alpha < t_k < \beta < t_{k+1}$ then for small $\varepsilon > 0$

$$y_\varepsilon(\alpha) > 0 \,, \quad y_\varepsilon(\beta) > 0 \,, \quad y_\varepsilon(t_k) = -\varepsilon < 0 \,,$$

and hence $y_\varepsilon(t)$ has at least two zeros in (t_{k-1}, t_{k+1}). Thus $y_\varepsilon(t)$ has at least n distinct zeros, which is a contradiction.

Suppose next that r_i is odd for $1 \leq i \leq h$. Then $r_k > 1$ for some k. Let $y_\varepsilon(t)$ be the nontrivial solution which satisfies the initial conditions $y_\varepsilon^{(j)}(t_k) = y^{(j)}(t_k)$ $(j = 0, \ldots, n-1;\ j \neq r_k-1)$, $y_\varepsilon^{(r_k-1)}(t_k) = \varepsilon \neq 0$. If ε is small then $y_\varepsilon(t)$ is close to $y(t)$ throughout the interval $[t_0, t_{h+1}]$. Since $y(t)$ changes sign at each of its zeros it follows that $y_\varepsilon(t)$ has $n - 2$ zeros near t_i $(i = 1, \ldots, h;\ i \neq k)$, besides a zero of multiplicity $r_k - 1$ at t_k. In the neighbourhood of t_k the function

$$\frac{y(t)}{y_\varepsilon(t)} = \frac{y^{(r_k)}(t_k)}{\varepsilon r_k}(t-t_k) + o(t-t_k)$$

changes sign and is close to 1 at α and β. Therefore $y_\varepsilon(t)$ has a zero in (t_{k-1}, t_{k+1}) distinct from t_k. Thus $y_\varepsilon(t)$ has at least n distinct zeros, which is a contradiction.

(ii) $h < n-1$.

The argument in the second part of (i) shows that if $h-1$ zeros of $y(t)$ are simple we can find another solution $y(t)$ with more than h distinct zeros of total multiplicity $\geq r \geq n$. This contradicts the definition of h.

Thus it only remains to consider the case where at least two zeros, t_k and t_m say, are not simple. Without loss of generality we can assume $y^{(r_k)}(t_k) > 0$. Choose integers $q_1, \ldots, q_h$ and $s \geq 1$ so that

$$1 \leq q_i \leq r_i \quad (i = 1, \ldots, h;\ i \neq k), \quad 0 \leq q_k = r_k - 2s,$$

and

$$q = q_1 + \ldots + q_h = n - 2.$$

There exists a unique solution $u(t)$ such that

$$u^{(j)}(t_i) = 0 \quad (i = 1, \ldots, h;\ i \neq k;\ j = 0, \ldots, q_i-1), \quad u(t_{h+1}) = 0,$$

$$u^{(j)}(t_k) = 0 \quad (j = 0, \ldots, q_k-1), \quad u^{(q_k)}(t_k) = 1,$$

since by the definition of h no nontrivial solution satisfies the corresponding homogeneous boundary conditions. The nontrivial solution $y_\varepsilon(t) = y(t) - \varepsilon u(t)$ has a zero at t_i of multiplicity at least q_i $(i = 1, \ldots, h;\ i \neq k)$. If $q_k \geq 1$ and $\varepsilon > 0$ it has a zero of order q_k at t_k. In the neighbourhood of t_k the function

$$\frac{y(t)}{y_\varepsilon(t)} = -\frac{y^{(r_k)}(t_k)}{\varepsilon r_k \ldots (q_k+1)} (t-t_k)^{2s} + o(t-t_k)^{2s}$$

is negative, whereas for small ε it is close to 1 at α and β. Hence $y_\varepsilon(t)$ has at least two zeros $\neq t_k$ in (t_{k-1}, t_{k+1}). Thus $y_\varepsilon(t)$ has at least $h+1$ distinct zeros of total multiplicity $\geq n$, which again contradicts the definition of h.

2. A test for disconjugacy

In this section we generalise Theorem 13 of Chapter 1 by showing that the equation (1) is disconjugate on a compact interval I if the coefficients $p_k(t)$ $(k = 1, \ldots, n)$ are sufficiently small relative to the length of I.

LEMMA 1. *Suppose* $y(t) \in C^{n-1}$ *has at least* n *zeros on* $I = [a, b]$. *Then we can find points* $a_0, a_1, \ldots, a_{2n-2}$ *such that*

$$a \le a_0 \le a_1 \le \ldots \le a_{2n-2} \le b$$

and

$$0 = y(a_0) = y'(a_1) = \ldots = y^{(n-2)}(a_{n-2}) = y^{(n-1)}(a_{n-1}) = y^{(n-2)}(a_n) = \ldots = y'(a_{2n-3}) = y(a_{2n-2}) .$$

Proof. Let $a_1^{(0)}, \ldots, a_n^{(0)}$ be n zeros of $y(t)$ such that $a \le a_1^{(0)} \le \ldots \le a_n^{(0)} \le b$. By Rolle's theorem we can find $n - 1$ zeros $a_1^{(1)}, \ldots, a_{n-1}^{(1)}$ of $y'(t)$ such that $a_i^{(0)} \le a_i^{(1)} \le a_{i+1}^{(0)}$ $(i = 1, \ldots, n-1)$. Repeating this process we obtain finally a zero $a_1^{(n-1)}$ of $y^{(n-1)}(t)$ between two zeros $a_1^{(n-2)}, a_2^{(n-2)}$ of $y^{(n-2)}(t)$. The points

$$a_1^{(0)}, a_1^{(1)}, \ldots, a_1^{(n-2)}, a_1^{(n-1)}, a_2^{(n-2)}, \ldots, a_{n-1}^{(1)}, a_n^{(0)}$$

satisfy the requirements of the lemma.

LEMMA 2. *Suppose* $y(t) \in C^n$ *on the interval* $I = [a, b]$ *and*

$$|y^{(n)}(t)| \le \mu \quad for \quad a \le t \le b ,$$

$$y(a_0) = y'(a_1) = \ldots = y^{(n-1)}(a_{n-1}) = 0 , \tag{2}$$

where $a \le a_0 \le a_1 \le \ldots \le a_n \le b$. *Then*

$$|y(t)| \le \frac{\mu(b-a)^n}{n[(n-1)/2]![n/2]!} \quad for \quad a \le t \le b . \tag{3}$$

Proof. From the identity

$$y^{(k)}(t) = \int_{a_k}^{t} \int_{a_{k+1}}^{t_1} \cdots \int_{a_{n-1}}^{t_{n-k-1}} y^{(n)}(t_{n-k})dt_{n-k}dt_{n-k-1} \ldots dt_1$$

we obtain for $t \le a_k$

$$|y^{(k)}(t)| \le \int_t^b \int_{t_1}^b \cdots \int_{t_{n-k-1}}^b \left|y^{(n)}(t_{n-k})\right| dt_{n-k} dt_{n-k-1} \cdots dt_1$$

$$\le \mu(b-t)^{n-k}/(n-k)! \ . \qquad (4)$$

In particular,

$$|y(t)| \le \mu(b-a)^n/n! \quad \text{for} \quad a \le t \le a_0 \ .$$

Similarly from

$$y(t) = \int_{a_0}^{t} \int_{a_1}^{t_1} \cdots \int_{a_{k-1}}^{t_{k-1}} \left|y^{(k)}(t_k)\right| dt_k dt_{k-1} \cdots dt_1$$

we obtain for $t \ge a_{k-1}$

$$|y(t)| \le \int_a^t \int_a^{t_1} \cdots \int_a^{t_{k-1}} \left|y^{(k)}(t_k)\right| dt_k dt_{k-1} \cdots dt_1$$

$$= \int_a^t |y^{(k)}(s)|(t-s)^{k-1} ds/(k-1)! \ . \qquad (5)$$

In particular

$$|y(t)| \le \mu \int_a^b (b-s)^{n-1} ds/(n-1)! = \mu(b-a)^n/n! \quad \text{for} \quad a_{n-1} \le t \le b \ .$$

From (4) and (5) we obtain for $a_{k-1} \le t \le a_k$

$$|y(t)| \le \mu \int_a^t (b-s)^{n-k}(t-s)^{k-1} ds/(n-k)!(k-1)!$$

$$\le \mu \int_a^b (b-s)^{n-1} ds/(n-k)!(k-1)!$$

$$= \frac{\mu(b-a)^n}{n(n-k)!(k-1)!} = \binom{n-1}{n-k}\mu(b-a)^n/n! \ .$$

When k runs through the values $1, \ldots, n$ the binomial coefficient $\binom{n-1}{n-k}$ takes its maximum value for $k = [(n+1)/2]$. It follows that

$$|y(t)| \le \frac{\mu(b-a)^n}{n[(n-1)/2]![n/2]!} \quad \text{for} \quad a \le t \le b \ .$$

By examining the proof it is easily seen that equality cannot hold in (3) unless $|y^{(n)}(t)| = \mu$ for $a \le t \le b$, *i.e.*, if $\mu > 0$, unless $y(t)$ is a polynomial of degree n .

THEOREM 1. *Suppose* $|p_k(t)| \le M_k$ *for all* t *in the compact interval* $I = [a, b]$ $(k = 1, \ldots, n)$. *Then the equation* (1) *is disconjugate on* I *if* $\chi\left(\frac{b-a}{2}\right) \le 1$, *where*

$$\chi(h) = \sum_{k=1}^{n} \frac{M_k h^k}{k[(k-1)/2]![k/2]!} .$$

Proof. Suppose on the contrary that (1) has a nontrivial solution $y(t)$ with at least n zeros. Then, by Lemma 1 with $a_{n-1} = c$, $y(t)$ satisfies the conditions (2) of Lemma 2 on a subinterval $[a, c]$, and the same conditions with the inequalities satisfied by $a_0, \ldots, a_{n-1}$ reversed on the complementary subinterval $[c, b]$. One of these two subintervals, say $[a, c]$, has length at most $(b-a)/2$. Moreover the interval $[a, c]$ is nondegenerate, since $y(t)$ cannot have a zero of multiplicity n . Applying Lemma 2 to this interval we obtain

$$\max_{a\le t\le c} |y^{(n-k)}(t)| \le \frac{\mu(b-a)^k}{2^k k[(k-1)/2]![k/2]!} \quad (k = 1, \ldots, n) , \tag{6}$$

where $\mu = \max_{a\le t\le c} |y^{(n)}(t)|$. But for some $\tau \in [a, c]$

$$\begin{aligned} \mu = |y^{(n)}(\tau)| &= \left|p_1(\tau)y^{(n-1)}(\tau) + \ldots + p_n(\tau)y(\tau)\right| \\ &\le \sum_{k=1}^{n} M_k |y^{(n-k)}(\tau)| \\ &\le \mu \sum_{k=1}^{n} \frac{M_k(b-a)^k}{2^k k[(k-1)/2]![k/2]!} . \end{aligned}$$

Evidently $\mu > 0$, since otherwise $y(t)$ would coincide on $[a, c]$ with a polynomial of degree $m < n$ and $y^{(m)}(t)$ would not vanish on $[a, c]$. Hence $\chi\left(\frac{b-a}{2}\right) \ge 1$. It only remains to exclude the possibility of equality. At least one of the numbers $M_1, \ldots, M_n$ is different from zero, since otherwise $y(t)$ would be a polynomial of degree less than n and could not have n zeros. Thus if $\chi\left(\frac{b-a}{2}\right) = 1$ then equality must hold in (6) for at least one value of k . This is possible only if $y(t)$ coincides on $[a, c]$ with a polynomial of degree n . But we can then take τ to be *any* point of $[a, c]$, and $|y^{(n-k)}(\tau)|$ is not constant on $[a, c]$ for any $k = 1, \ldots, n$. Therefore also in this case we have $\chi\left(\frac{b-a}{2}\right) > 1$.

3. Markov and Descartes systems

The functions $y_1(t), \ldots, y_n(t) \in C^n$ are said to form a *Čebyšev system* on the interval I if every nontrivial linear combination of $y_1(t), \ldots, y_n(t)$ has less than n zeros. They are said to form a *Markov system* if the n Wronskians

$$W(y_1, \ldots, y_k) = \begin{vmatrix} y_1 & \cdots & y_k \\ & \cdots\cdots & \\ y_1^{(k-1)} & \cdots & y_k^{(k-1)} \end{vmatrix} \qquad (k = 1, \ldots, n)$$

are positive throughout I. They are said to form a *Descartes system* if all ordered Wronskians

$$W\left(y_{i_1}, \ldots, y_{i_k}\right) = \begin{vmatrix} y_{i_1} & \cdots & y_{i_k} \\ & \cdots\cdots & \\ y_{i_1}^{(k-1)} & \cdots & y_{i_k}^{(k-1)} \end{vmatrix} \qquad \left(1 \le i_1 < \ldots < i_k \le n;\ k = 1, \ldots, n\right)$$

are positive throughout I.

It is evident from the definition that the equation (1) is disconjugate on an interval I if and only if some, and then, every fundamental system is a Čebyšev system. Before studying the connection of Markov and Descartes systems with questions of disconjugacy we will derive some formal properties of Wronskians.

LEMMA 3. *If* $r(t) \neq 0$ *then*

$$W\left(ry_1, \ldots, ry_k\right) = r^k W\left(y_1, \ldots, y_k\right) .$$

In particular, if $y_1 \neq 0$ *and* $z_i = \left(y_{i+1}/y_1\right)'$ $(i = 1, \ldots, k-1)$ *then*

$$W\left(y_1, y_2, \ldots, y_k\right) = y_1^k W\left(z_1, \ldots, z_{k-1}\right) .$$

Proof. The first statement follows from the fact that

$$(ru)^{(i)} = ru^{(i)} + R_i ,$$

where R_i is a linear combination of $u, u', \ldots, u^{(i-1)}$ with coefficients depending only on r. The second statement is obtained by taking $r = y_1^{-1}$.

LEMMA 4. *If* $W\left(y_1, \ldots, y_{k-1}\right) \neq 0$ *and* $W\left(y_1, \ldots, y_k\right) \neq 0$ *on* I *then*

$$\left[\frac{W\left(y_1, \ldots, y_{k-1}, y\right)}{W\left(y_1, \ldots, y_k\right)}\right]' = \frac{W\left(y_1, \ldots, y_{k-1}\right) W\left(y_1, \ldots, y_k, y\right)}{\left[W\left(y_1, \ldots, y_k\right)\right]^2} .$$

Proof. If we set the left side equal to zero we obtain a linear differential equation of order k with leading coefficient $W(y_1, \ldots, y_{k-1})/W(y_1, \ldots, y_k)$ which has the linearly independent solutions $y_1, \ldots, y_k$. Similarly if we set the right side equal to zero we obtain a linear differential equation of order k with the same leading coefficient and the same solutions. The result follows.

In the same way we can prove

LEMMA 5. *If* $W(y_1, \ldots, y_{k-1}) \neq 0$ *and* $W(y_2, \ldots, y_k) \neq 0$ *on* I *then*

$$W(y_2, \ldots, y_k)W(y_1, \ldots, y_{k-1}, y) = W(y_1, \ldots, y_{k-1})W(y_2, \ldots, y_k, y) + W(y_2, \ldots, y_{k-1}, y)W(y_1, \ldots, y_k) .$$

We will use this last result to show that the number of Wronskians which are assumed positive in the definition of a Descartes system can be substantially reduced.

PROPOSITION 4. *The functions* $y_1, \ldots, y_n$ *form a Descartes system on the interval* I *if all consecutive Wronskians*

$$W(y_i, y_{i+1}, \ldots, y_j) \quad (1 \leq i \leq j \leq n)$$

are positive on I .

Proof. If we use induction on n then

$$W\left(y_{i_1}, \ldots, y_{i_k}\right) > 0 \quad \text{for } 1 < i_1 < \ldots < i_k \leq n .$$

We have to show that the same holds for $i_1 = 1$. This is certainly true for $k = 1$. We assume that $k > 1$ and the result holds for all smaller values of k. We suppose the sets of indices $(i_2, \ldots, i_k)$ arranged in lexicographic order. Then the first set is $(2, \ldots, k)$ and $W(y_1, y_2, \ldots, y_k) > 0$. We assume that $i_k > k$ and the result holds for all preceding sets. Let j be the least integer such that $i_j > j$. It follows from Lemma 5 that

$$W\left(y_2, \ldots, y_j, y_{i_j}, \ldots, y_{i_{k-1}}\right)W\left(y_{i_1}, y_{i_2}, \ldots, y_{i_{k-1}}, y\right) = W\left(y_{i_1}, y_{i_2}, \ldots, y_{i_{k-1}}\right)W\left(y_2, \ldots, y_j, y_{i_j}, \ldots, y_{i_{k-1}}, y\right) + W\left(y_{i_2}, \ldots, y_{i_{k-1}}, y\right)W\left(y_1, \ldots, y_j, y_{i_j}, \ldots, y_{i_{k-1}}\right) .$$

Substituting y_{i_k} for y we obtain from the various induction hypotheses

$W\left(y_{i_1}, \ldots, y_{i_k}\right) > 0$.

The significance of Markov systems is brought out by the following result. It shows that if $y_1, \ldots, y_n$ is a Markov system then $y_1, \ldots, y_k$ is a Čebyšev system for $k = 1, \ldots, n$.

PROPOSITION 5. *Let* $y_1(t), \ldots, y_n(t)$ *be functions in* $C^n(I)$. *Then the number of zeros of an arbitrary nontrivial linear combination* $\alpha_1 y_1 + \ldots + \alpha_k y_k$ *is less than* k *for* $k = 1, \ldots, n$ *if and only if the Wronskians* $W(y_1, \ldots, y_k)$ *do not vanish on* I *for* $k = 1, \ldots, n$.

Proof. In one direction this is immediate, since if $W(y_1, \ldots, y_k)$ vanishes at some point c there is a nontrivial linear combination of $y_1, \ldots, y_k$ which has a zero at c of multiplicity at least k . Suppose on the other hand that the n Wronskians do not vanish, so that in particular $y_1 \neq 0$. Since the result obviously holds for $n = 1$ we assume that $n > 1$ and that it holds for systems of less than n functions. If a linear combination $y = \alpha_1 y_1 + \ldots + \alpha_n y_n$ has at least n zeros then $(y/y_1)'$ has at least $n - 1$ zeros, by Rolle's theorem between two consecutive zeros and by Leibniz's product formula at a multiple zero. But if we put $z_k = (y_{k+1}/y_1)'$ $(k = 1, \ldots, n-1)$ then $(y/y_1)' = \alpha_2 z_1 + \ldots + \alpha_n z_{n-1}$ and, by Lemma 3, $W(z_1, \ldots, z_k) \neq 0$ $(k = 1, \ldots, n-1)$. Therefore, by the induction hypothesis, $\alpha_2 = \ldots = \alpha_n = 0$. Hence $y = \alpha_1 y_1$, which vanishes only if $\alpha_1 = 0$ also.

The significance of Descartes systems is brought out by the following result. It shows, in particular, that if $y_1, \ldots, y_n$ is a Descartes system then any subset is a Čebyšev system.

PROPOSITION 6. *Let* $y_1(t), \ldots, y_n(t)$ *be functions in* $C^n(I)$. *Then the number of zeros of an arbitrary nontrivial linear combination* $y = \alpha_1 y_1 + \ldots + \alpha_n y_n$ *is at most equal to the number of sign changes in the sequence of coefficients* $\alpha_1, \ldots, \alpha_n$ *(after omitting all zero terms) if and only if no Wronskian* $W\left(y_{i_1}, \ldots, y_{i_k}\right)$, *where* $1 \leq i_1 < \ldots < i_k \leq n$, *vanishes on* I *and any two Wronskians corresponding to the same value of* k *have the same sign.*

Proof. Suppose first that the number of zeros is at most equal to the number of sign changes. If the Wronskian $W\left(y_{i_1}, \ldots, y_{i_k}\right)$ vanished at c some nontrivial linear combination $y = \alpha_1 y_{i_1} + \ldots + \alpha_k y_{i_k}$ would have a zero of multiplicity $\geq k$ at c, whereas the sequence of coefficients has at most $k - 1$ sign changes. Therefore $W\left(y_{i_1}, \ldots, y_{i_k}\right) \neq 0$. Thus there exist constants $\alpha_1, \ldots, \alpha_k$ such that

$$\alpha_1 y_{i_1}(c) + \alpha_2 y_{i_2}(c) + \ldots + \alpha_k y_{i_k}(c) = 0$$

$$\cdots\cdots$$

$$\alpha_1 y_{i_1}^{(k-2)}(c) + \alpha_2 y_{i_2}^{(k-2)}(c) + \ldots + \alpha_k y_{i_k}^{(k-2)}(c) = 0$$

$$\alpha_1 y_{i_1}^{(k-1)}(c) + \alpha_2 y_{i_2}^{(k-1)}(c) + \ldots + \alpha_k y_{i_k}^{(k-1)}(c) = 1 .$$

Then $y = \alpha_1 y_{i_1} + \ldots + \alpha_k y_{i_k}$ has a zero of multiplicity $k - 1$ at c. Therefore the sequence of coefficients $\alpha_1, \ldots, \alpha_k$ has exactly $k - 1$ sign changes. But

$$\alpha_j W\left(y_{i_1}, \ldots, y_{i_k}\right) = (-1)^{j+k} W\left(y_{i_1}, \ldots, \hat{y}_{i_j}, \ldots, y_{i_k}\right),$$

where $\hat{y}_{i_g}$ denotes that y_{i_g} is omitted.

Therefore the Wronskians obtained from $W\left(y_{i_1}, \ldots, y_{i_k}\right)$ by omitting one function all have the same sign. Moreover, any two Wronskians of $k - 1$ functions can be obtained from one another by repeatedly changing one function at a time.

Suppose next that the Wronskian condition is satisfied. Since the result is trivial for $n = 1$ we assume that $n > 1$ and that it holds for systems of less then n functions. The functions $y_1, \ldots, y_n$ all have the same sign on I and without loss of generality we can assume that they are positive. Let $y = \alpha_1 y_1 + \ldots + \alpha_n y_n$ be any nontrivial linear combination of them with V sign changes in the sequence of coefficients. Clearly y has no zeros if $\alpha_1, \ldots, \alpha_n$ all have the same sign, *i.e.*, if $V = 0$. Hence we can assume that there is at least one sign change, involving α_h say. Put

$$z_i = \begin{cases} -\left(y_i/y_h\right)' & \text{for } 1 \leq i < h , \\ \left(y_{i+1}/y_h\right)' & \text{for } h \leq i < n . \end{cases}$$

Then $z_1, \ldots, z_{n-1}$ satisfy the same conditions as $y_1, \ldots, y_n$, since if $1 \leq i_1 < \ldots < i_j < h < i_{j+1} < \ldots < i_k \leq n$,

$$W\left(y_{i_1}, \ldots, y_{i_j}, y_h, y_{i_{j+1}}, \ldots, y_{i_k}\right) = y_h^{k+1} W\left(z_{i_1}, \ldots, z_{i_j}, z_{i_{j+1}}, \ldots, z_{i_k}\right) .$$

Since

$$z \equiv (y/y_h)' = -\alpha_1 z_1 - \ldots - \alpha_{h-1} z_{h-1} + \alpha_{h+1} z_h + \ldots + \alpha_n z_{n-1} ,$$

the number of sign changes in the sequence of coefficients for z is $V - 1$. Hence, by the induction hypothesis, z has at most $V - 1$ zeros. It follows that y has at most V zeros.

For example, if $\lambda_1 < \ldots < \lambda_n$, we have

$$W\left(e^{\lambda_1 t}, \ldots, e^{\lambda_k t}\right) = e^{\lambda_1 t} \ldots e^{\lambda_k t} \begin{vmatrix} 1 & \ldots & 1 \\ \lambda_1 & \ldots & \lambda_k \\ & \ldots\ldots & \\ \lambda_1^{k-1} & \ldots & \lambda_k^{k-1} \end{vmatrix}$$

$$= e^{(\lambda_1 + \ldots + \lambda_k)t} \prod_{1 \le i < j \le k} (\lambda_j - \lambda_i)$$

$$> 0 .$$

Hence if $p(\lambda)$ is a polynomial in λ with constant coefficients and real distinct zeros $\lambda_1, \ldots, \lambda_n$, where $\lambda_1 < \ldots < \lambda_n$, the fundamental system of solutions $e^{\lambda_1 t}, \ldots, e^{\lambda_n t}$ of the autonomous differential equation $p(D)y = 0$ is a Descartes system on any real interval. Thus the number of zeros of any linear combination $y(t) = \alpha_1 e^{\lambda_1 t} + \ldots + \alpha_n e^{\lambda_n t}$ is at most equal to the number of sign changes in the sequence of coefficients $\alpha_1, \ldots, \alpha_n$. If we take $\lambda_k = k - 1$ $(k = 1, \ldots, n)$ and put $x = e^t$ we obtain Descartes' rule of signs for the number of positive zeros of a polynomial.

It is time now to consider the connection between Markov and Descartes systems and differential equations.

THEOREM 2. *The linear differential equation* (1) *has a Markov fundamental system of solutions if and only if the operator* L *has a representation*

$$Ly \equiv v_1 v_2 \ldots v_n D \frac{1}{v_n} D \ldots D \frac{1}{v_2} D \frac{1}{v_1} y , \tag{7}$$

where $D = d/dt$, $v_k > 0$ *and* $v_k \in C^{n-k+1}$ $(k = 1, \ldots, n)$.

Proof. Let $y_1, \ldots, y_n$ be a Markov system, so that

$$W_k = W(y_1, \ldots, y_k) > 0 \quad \text{for} \quad k = 1, \ldots, n .$$

The differential equation with leading coefficient 1 which has $y_1, \ldots, y_k$ as a fundamental system of solutions is given by

$$L_k y \equiv W(y_1, \ldots, y_k, y)/W_k = 0 .$$

Put

$$1 = W_0 , \quad v_1 = W_1 , \quad v_k = W_k W_{k-2}/W_{k-1}^2 \quad (k = 2, \ldots, n) ,$$

so that

$$W_k/W_{k-1} = v_1 v_2 \ldots v_k .$$

We will prove by induction that

$$L_k y \equiv v_1 \ldots v_k D \frac{1}{v_k} D \ldots D \frac{1}{v_1} y .$$

This is trivial for $k = 1$. Suppose it holds as written. Then

$$\begin{aligned} v_1 \ldots v_{k+1} D \frac{1}{v_{k+1}} D \frac{1}{v_k} D \ldots D \frac{1}{v_1} y &= v_1 \ldots v_k D \frac{1}{v_1 \ldots v_k v_{k+1}} L_{k+1} y \\ &= \frac{W_{k+1}}{W_k} D \left(\frac{W_k}{W_{k+1}} L_k y \right) \\ &= \frac{W_{k+1}}{W_k} D \left(\frac{W(y_1, \ldots, y_k, y)}{W_{k+1}} \right) . \end{aligned}$$

By Lemma 4 the right side is equal to $W(y_1, \ldots, y_{k+1}, y)/W_{k+1}$.

Conversely, suppose L has a representation (7). Define $y_1, \ldots, y_n$ inductively by setting $y_1 = v_1$ and taking y_k to be a solution of the equation

$$W(y_1, \ldots, y_{k-1}, y)/W_{k-1} = v_1 v_2 \ldots v_k \quad (k = 2, \ldots, n) .$$

Then $y_1, \ldots, y_n$ are a Markov system and the above argument can be reversed to yield

$$Ly \equiv W(y_1, \ldots, y_n, y)/W_n .$$

Thus, $y_1, \ldots, y_n$ are a fundamental system of solutions of (1).

It is easily verified that if $v_k > 0$ and $v_k \in C^{n-k+1}$ then

$$(D-r_n) \ldots (D-r_1) y \equiv v_1 \ldots v_n D \frac{1}{v_n} D \ldots D \frac{1}{v_1} y , \tag{8}$$

where

$$r_k = (v_1'/v_1) + \dots + (v_k'/v_k) \quad (k = 1, \dots, n) .$$

Conversely for any functions $r_k \in C^{n-k}$, (8) holds with

$$v_k(t) = \exp \int^t [r_k(s)-r_{k-1}(s)]ds \quad (k = 1, \dots, n;\ r_0 \equiv 0) .$$

Thus the representation (7) is equivalent to a factorisation of L into linear factors.

If the equation (1) has a Markov fundamental system of solutions on an interval I , then it is disconjugate on I , since any Markov system is a Čebyšev system. We now consider to what extent the converse holds.

LEMMA 6. *If the equation* (1) *is disconjugate on an interval* I *and* $a \in I$, *then it has a Markov fundamental system of solutions for* $t > a$.

Proof. Let $y_1, \dots, y_n$ be solutions of (1) satisfying initial conditions of the form

$$y_k = y_k' = \dots = y_k^{(n-k-1)} = 0 , \quad y_k^{(n-k)} \neq 0 \quad \text{at} \quad t = a .$$

The solutions $y_1, \dots, y_n$ are evidently linearly independent. If the Wronskian $W(y_1, \dots, y_k)$ vanished at some point $c > a$ then a nontrivial linear combination of $y_1, \dots, y_k$ would have a zero at c of multiplicity $\geq k$. Since any such linear combination has a zero at a of multiplicity $\geq n - k$ this contradicts the disconjugacy of (1). By replacing y_k by $- y_k$ if necessary we can ensure that $W(y_1, \dots, y_k) > 0$ for $t > a$ and $k = 1, \dots, n$.

LEMMA 7. *Let* $I = [a, b]$ *be a compact interval. If the equation* (1) *is disconjugate on a proper subinterval* $[a_1, b_1]$ *then it is also disconjugate on the interval* $[a_1-\varepsilon, b_1+\varepsilon]$ *for sufficiently small* $\varepsilon > 0$.

Proof. Otherwise for every large integer ν there is a nontrivial solution $y_\nu(t)$ of (1) with n zeros $t_{\nu,1}, \dots, t_{\nu,n}$ on the interval $[a_1-1/\nu, b_1+1/\nu]$. We can suppose the solution normalised so that

$$[y_\nu(b)]^2 + [y_\nu'(b)]^2 + \dots + \left[y_\nu^{(n-1)}(b)\right]^2 = 1 .$$

By restricting attention to a suitable subsequence we obtain a nontrivial solution $y(t)$ and points $t_1, \dots, t_n$ in $[a_1, b_1]$ such that, as $\nu \to \infty$,

$y_\nu^{(k-1)}(t) \to y^{(k-1)}(t)$ uniformly on I and $t_{\nu,k} \to t_k$ $(k = 1, \ldots, n)$. It is easily seen that $y(t)$ has at least n zeros on $[a_1, b_1]$, even if some of the points t_k are repeated. Thus we have a contradiction.

We can now prove

THEOREM 3. *The equation* (1) *has a Markov fundamental system of solutions on the compact interval* $I = [a, b]$ *if, and only if, it is disconjugate on* I.

Proof. By setting $p_k(t) = p_k(a)$ for $t < a$ we can extend the domain of definition of the coefficients of (1) to the interval $(-\infty, b]$. By Lemma 7 the equation (1) is disconjugate on the interval $[c, b]$ if c is less than a and sufficiently close to a. By Lemma 6 the equation (1) has a Markov fundamental system of solutions on the interval $(c, b]$, and therefore on $[a, b]$.

This result can be used to derive several properties of disconjugate equations.

PROPOSITION 7. *Suppose the equation* (1) *is disconjugate on the compact interval* $I = [a, b]$. *If* $y \in C^n$ *has* $n + 1$ *zeros, not all coincident, on* I *then* Ly *has at least one zero in the open interval* (a, b).

Proof. By Theorem 3 the equation (1) has a Markov fundamental system on I. Hence, by Theorem 2, the operator L has a representation of the form (7). If $n = 1$ the result now follows from Rolle's theorem. Suppose $n > 1$ and the result holds for equations of order less than n. Then, by Rolle's theorem again, $D \frac{1}{v_1} y$ has at least n zeros, not all coincident, in I. Hence, by the induction hypothesis, Ly has at least one zero in (a, b).

PROPOSITION 8. *If the equations* $L_1 y = 0$, $L_2 y = 0$ *are disconjugate on the interval* I *then the composite equation* $L_1(L_2 y) = 0$ *is also disconjugate on* I.

Proof. It is sufficient to prove the proposition for the case when I is compact. Then, by Theorem 3, both equations have Markov fundamental systems. Hence, by Theorem 2, the linear operators L_1, L_2 have representations of the form (7). The composite operator $L_1 L_2$ likewise has such a representation and consequently, by Theorem 2 again, the equation $L_1 L_2 y = 0$ is disconjugate on I.

We can make the set of all equations (1) on a compact interval I into a metric space by defining the distance between two equations $(1)_1$ and $(1)_2$ to be

$$\sup_{t \in I} \sum_{k=1}^{n} |p_{k,1}(t) - p_{k,2}(t)| .$$

Then we have

PROPOSITION 9. *The set of all disconjugate equations* (1) *on a compact interval* I *is connected and open.*

Proof. If the equation (1) is disconjugate on I then by Theorem 3 and Theorem 2 the operator L has a representation

$$Ly \equiv v_{n+1} D \frac{1}{v_n} D \dots D \frac{1}{v_1} y ,$$

where $D = d/dt$, $v_k > 0$ and $v_k \in C^{n-k+1}$ $(k = 1, \dots, n)$, and $v_{n+1} = v_1 \dots v_n$. The operator L can be connected to the operator D^n by the path

$$L^{(\lambda)} = v_{n+1}^{\lambda} D \frac{1}{v_n^{\lambda}} D \dots D \frac{1}{v_1^{\lambda}} \quad (0 \le \lambda \le 1)$$

and, by Theorem 2, the equations $L^{(\lambda)}y = 0$ are all disconjugate.

Suppose that for each positive integer ν there is an equation $(1)_\nu$, distant less than $1/\nu$ from (1), which is not disconjugate. Let $y_\nu(t)$ be a solution of $(1)_\nu$ with at least n zeros such that $y_\nu^2 + y_\nu'^2 + \dots + y_\nu^{(n-1)^2} = 1$ at $t = a$. By restricting attention to a suitable subsequence we can assume that $y_\nu^{(k-1)}(a) \to \eta_{k-1}$ as $\nu \to \infty$ $(k = 1, \dots, n)$. Then $y_\nu^{(k-1)}(t) \to y^{(k-1)}(t)$ uniformly on I as $\nu \to \infty$ $(k = 1, \dots, n)$, where $y(t)$ is the nontrivial solution of (1) which satisfies the initial conditions $y^{(k-1)}(a) = \eta_{k-1}$. It is easily seen that $y(t)$ has at least n zeros on I . From this contradiction we conclude that there exists $\delta > 0$ such that every equation distant less than δ from (1) is also disconjugate.

Theorem 1 provides a quantitative estimate for δ in the case of the equation $y^{(n)} = 0$.

We turn now to Descartes systems.

LEMMA 8. *If* $y_1, \dots, y_n$ *is a Markov system on an interval* I *such that, for some* a *in* I , y_k *has a zero at* a *of multiplicity* $k - 1$ $(k = 1, \dots, n)$ *then* $y_1, \dots, y_n$ *is a Descartes system for* $t > a$.

Proof. We use induction on n . Then $y_1, \dots, y_{n-1}$ form a Descartes system for $t > a$. To prove that $y_1, \dots, y_n$ also form a Descartes system it is sufficient, by Proposition 4, to show that for $t > a$

$$W(y_k, y_{k+1}, \ldots, y_n) > 0 \quad (k = 1, \ldots, n) .$$

This is certainly true for $k = 1$. We assume that $k > 1$ and it holds for all smaller values of k. Then, by Lemma 4,

$$\left[\frac{W(y_k,\ldots,y_n)}{W(y_{k-1},\ldots,y_{n-1})}\right]' = \frac{W(y_k,\ldots,y_{n-1})W(y_{k-1},\ldots,y_n)}{[W(y_{k-1},\ldots,y_{n-1})]^2} > 0 \quad \text{for} \quad t > a .$$

Hence $W(y_k, \ldots, y_n)$ will certainly be positive for $t > a$ if it vanishes at $t = a$. But this is the case for $k > 1$, because y_n has a zero at a of multiplicity $n - 1$.

LEMMA 9. *If (1) has a Markov fundamental system of solutions on an interval* I *then, for any* a *in* I, *(1) has a Markov fundamental system of solutions* $y_1, \ldots, y_n$ *such that* y_k *has a zero at* a *of multiplicity exactly* $k - 1$ $(k = 1, \ldots, n)$.

Proof. Let $z_1, \ldots, z_n$ be the given Markov system. We define $y_1, \ldots, y_n$ by setting $y_1(t) = z_1(t)$ and

$$y_k(t) = \begin{vmatrix} z_1(a) & \ldots & z_k(a) \\ & \ldots\ldots & \\ z_1^{(k-2)}(a) & \ldots & z_k^{(k-2)}(a) \\ z_1(t) & \ldots & z_k(t) \end{vmatrix} \quad \text{for} \quad 2 \le k \le n .$$

Thus

$$y_k(t) = \sum_{j=1}^{k} (-1)^{k-j}\beta_{kj}z_j(t) ,$$

where $\beta_{kj} = W(z_1, \ldots, \hat{z}_j, \ldots, z_k)(a)$. In particular $\beta_{kk} = W(z_1, \ldots, z_{k-1})(a) > 0$. It follows that

$$y_k(a) = \ldots = y_k^{(k-2)}(a) = 0 , \quad y_k^{(k-1)}(a) = W(z_1, \ldots, z_k)(a) > 0 .$$

Moreover $y_1, \ldots, y_n$ is a Markov system on I, since

$$W(y_1) = z_1 > 0 , \quad W(y_1, \ldots, y_k) = \beta_{22} \ldots \beta_{kk}W(z_1, \ldots, z_k) > 0 \quad (k > 1) .$$

This construction may also be applied if $z_1, \ldots, z_n$ are not solutions of the differential equation (1) but of corresponding differential inequalities. Let $z_1, \ldots, z_n$ be a Markov system such that $(-1)^{n-k}L[z_k] \ge 0$ $(k = 1, \ldots, n)$ and

$$W(z_1, \dots, \hat{z}_j, \dots, z_k) > 0 \quad \text{at} \quad t = a \quad \text{for} \quad 1 \le j < k \le n .$$

Then $y_1, \dots, y_n$ is a Markov system such that y_k has a zero at a of multiplicity exactly $k - 1$ and $(-1)^{n-k} L[y_k] \ge 0$ $(k = 1, \dots, n)$.

Combining Lemmas 8 and 9 we obtain

LEMMA 10. *If* (1) *has a Markov fundamental system of solutions on an interval* I *and* $a \in I$ *then* (1) *has a Descartes fundamental system of solutions for* $t > a$.

The existence of a Markov fundamental system of solutions of (1) is equivalent to a factorisation of L of the form (7). If L is given by (7) then a Markov fundamental system $y_1, \dots, y_n$ such that y_k has a zero at a of multiplicity $k - 1$ is given explicitly by

$$y_1(t) = v_1(t)$$

$$y_2(t) = v_1(t) \int_a^t v_2(t_1) dt_1$$

.....

$$y_n(t) = v_1(t) \int_a^t v_2(t_1) \dots \int_a^{t_{n-2}} v_n(t_{n-1}) dt_{n-1} \dots dt_1 .$$

In fact $W(y_1, \dots, y_k) = v_1^k v_2^{k-1} \dots v_k$ $(k = 1, \dots, n)$. Consequently the functions $y_1, \dots, y_n$ thus defined form a Descartes system for $t > a$.

THEOREM 4. *If the equation* (1) *has a Markov fundamental system of solutions on an interval* $I = [a, b]$ *or* $[a, b)$ *then it has a Descartes fundamental system of solutions on* I .

Proof. Extend the domain of definition of the coefficients of (1) by setting $p_k(t) = p_k(a)$ for $t < a$. If $y_1, \dots, y_n$ are a Markov fundamental system of solutions on I then the inequalities $W(y_1, \dots, y_k) > 0$ $(k = 1, \dots, n)$ continue to hold on an interval $J = [c, b]$ or $[c, b)$, where c is less than a and sufficiently close to a . Therefore, by Lemma 10, the equation (1) has a Descartes fundamental system of solutions on I .

Combining Theorems 3 and 4 we obtain

THEOREM 5. *The equation* (1) *has a Descartes fundamental system of solutions on a compact interval* $I = [a, b]$ *if, and only if, it is disconjugate on* I .

By combining the proofs of Theorems 3 and 4 we can actually construct a Descartes fundamental system of solutions. Extend the domain of definition of the coefficients of (1) by setting $p_k(t) = p_k(a)$ for $t < a$ and $= p_k(b)$ for $t > b$. Then (1) remains disconjugate on some interval $J = [a_1, b_1]$, where $a_1 < a < b < b_1$. Let $y_k(t)$ be the uniquely determined solution of (1) which satisfies the boundary conditions

$$y = y' = \dots = y^{(k-2)} = 0 \ , \quad y^{(k-1)} = 1 \quad \text{at} \quad t = a_1 \ ,$$

$$y = y' = \dots = y^{(n-k-1)} = 0 \qquad\qquad \text{at} \quad t = b_1 \ .$$

We will show that $y_1, \dots, y_n$ form a Descartes system on the open interval (a_1, b_1). Since y_k has a zero of multiplicity $k - 1$ at a_1 $(k = 1, \dots, n)$ it is sufficient to show that $y_1, \dots, y_n$ form a Markov system on $[a_1, b_1)$. Since any nontrivial linear combination of $y_1, \dots, y_k$ has a zero of multiplicity $\geq n - k$ at b_1 it cannot have a zero of multiplicity $\geq k$ at $t < b_1$ and hence the Wronskian $W(y_1, \dots, y_k)$ cannot vanish for $t < b_1$. Since $W(y_1, \dots, y_k) = 1$ at a_1 it follows that $W(y_1, \dots, y_k) > 0$ for $a_1 \leq t < b_1$ $(k = 1, \dots, n)$. This completes the proof.

4. Conjugate points

Throughout this section the interval I will be *open*. As a special case of Lemma 7 we have

LEMMA 11. *For any* a *in* I *there exists* $\delta = \delta(a) > 0$ *such that the equation* (1) *is disconjugate on the subinterval* $[a-\delta, a+\delta]$.

We will say that a nontrivial solution $y(t)$ of (1) has the *property* (k, a, b), where $1 \leq k \leq n-1$, if it has a zero of multiplicity $\geq k$ at b and a zero of multiplicity $\geq n - k$ at a.

Let $y_k(t, a)$ be the solution of (1) which satisfies at $t = a$ the initial conditions

$$y_k^{(n-k)} = 1 \ , \quad y_k^{(n-j)} = 0 \quad (j = 1, \dots, n; \ j \neq k) \ .$$

We denote by

$$W_k(t, a) = \begin{vmatrix} y_1(t, a) & \dots & y_k(t, a) \\ y_1'(t, a) & \dots & y_k'(t, a) \\ & \dots\dots & \\ y_1^{(k-1)}(t, a) & \dots & y_k^{(k-1)}(t, a) \end{vmatrix} \tag{9}$$

the Wronskian of $y_1(t, a), \dots, y_k(t, a)$. By the continuous dependence of solutions on initial values W_k is a continuous function of (t, a) on $I \times I$.

LEMMA 12. *There exists a solution with the property* (k, a, b) *if and only if* $W_k(b, a) = 0$.

In fact a solution has a zero of multiplicity $\geq n - k$ at a if and only if it is a linear combination of $y_1(t, a), \dots, y_k(t, a)$. Such a linear combination has a zero of multiplicity $\geq k$ at b if and only if $W_k(b, a) = 0$.

Suppose now that the equation (1) is not disconjugate on I. Then for some a in I there exists $b > a$ in I such that (1) is not disconjugate on $[a, b]$. Moreover if a has this property, so does any $a' < a$ in I. We denote by $\eta_+(a) = \eta(a)$ the supremum of all $c > a$ such that (1) is disconjugate on $[a, c]$. Thus if $\eta(a)$ is defined, so is $\eta(a')$ for any $a' < a$ in I and $\eta(a') \leq \eta(a)$. We call $\eta(a)$ the first right *conjugate point* of a. Also let $\omega_+(a) = \omega(a)$ be the least $b > a$ in I, if one exists, at which one of the Wronskians $W_1(b, a), \dots, W_{n-1}(b, a)$ vanishes. It may be noted that $W_n(b, a)$ never vanishes because the solutions $y_1, \dots, y_n$ are linearly independent.

PROPOSITION 10. $\eta(a) = \omega(a)$.

Proof. By Lemma 12 and the definition of $\eta(a)$ we have $\eta(a) \leq \omega(a)$. Suppose $\eta(a) < \omega(a)$ and choose c in I so that $\eta(a) < c < \omega(a)$. By Lemma 11 the equation (1) is disconjugate on an interval $[a-\delta, a+\delta]$, where $\delta > 0$. Since the functions W_k are continuous and $W_k(t, a) \neq 0$ for $a < t < \omega(a)$ we can choose $\delta_0 < \delta/2$ so that $W_k(t, s) \neq 0$ for $a-\delta_0 \leq s \leq a+\delta_0$, $a+\delta/2 \leq t \leq c$ and $k = 1, \dots, n-1$. By Lemma 12 we cannot have $W_k(t, a-\delta_0) = 0$ for some t in $(a-\delta_0, a+\delta/2)$ and some k, because (1) is disconjugate on $[a-\delta, a+\delta]$. Thus $W_k(t, a-\delta_0) \neq 0$ for $a-\delta_0 < t \leq c$. Hence the equation (1) has a Markov fundamental system of solutions, and consequently is disconjugate, on the interval $(a-\delta_0, c]$. But this implies $\eta(a) \geq c$, which is a contradiction.

PROPOSITION 11. *Let* $b = \eta(a)$ *and let* k *be the least integer* i *such that*

$W_i(b, a) = 0$. *The corresponding solution* $y(t)$ *of* (1) *with the property* (k, a, b) *is uniquely determined up to a constant factor, and does not vanish on the open interval* (a, b) .

Proof. We must have $y^{(n-k)}(a) \neq 0$, since otherwise the solution $y(t)$ would have the property $(k-1, a, b)$. If another solution $z(t)$ has the property (k, a, b) then

$$u(t) = y^{(n-k)}(a)z(t) - z^{(n-k)}(a)y(t)$$

is either identically zero or has the property $(k-1, a, b)$. Since the latter is excluded by the minimal nature of k it follows that $y(t)$ and $z(t)$ are linearly dependent.

Since $y(t)$ has a zero of multiplicity $\geq n - k$ at a we can write

$$y(t) = \alpha_1 y_1(t, a) + \dots + \alpha_k y_k(t, a) .$$

Moreover $\alpha_k \neq 0$, since $W_{k-1}(b, a) \neq 0$ and $y(t)$ has a zero of multiplicity $\geq k$ at b . Suppose $y(c) = 0$ for some c in (a, b) . The differential equation

$$L_{k-1}y \equiv \begin{vmatrix} y_1(t, a) & \dots & y_{k-1}(t, a) & y \\ & \dots\dots & & \\ y_1^{(k-1)}(t, a) & \dots & y_{k-1}^{(k-1)}(t, a) & y^{(k-1)} \end{vmatrix} = 0$$

has the solutions $y_1(t), \dots, y_{k-1}(t)$. By Theorem 2 the operator L_{k-1} has a factorisation of the form (7) on the interval $[c, b]$. Since $y(t)$ has a zero at c and a zero of multiplicity $\geq k$ at b it follows from Proposition 7 that $L_{k-1}y$ vanishes at some point c' in (c, b) . Replacing $y(t)$ by its expression as a linear combination of $y_1(t, a), \dots, y_k(t, a)$ we conclude that $W_k(c', a) = 0$. But this contradicts the definition of $b = \eta(a)$.

THEOREM 6. $\eta(a)$ *is an increasing function of* a .

Proof. We already know that $\eta(a)$ is a non-decreasing function. Suppose $a_1 < a_2$ and $\eta(a_1) = \eta(a_2) = b$. Then $\eta(t) = b$ for $a_1 \leq t \leq a_2$ and for each such t there is an integer k such that $W_k(t, b) = 0$. Let S_k denote the set of all t in $[a_1, a_2]$, if any, such that $W_k(t, b) = 0$ $(k = 1, \dots, n-1)$. Then S_k is closed and the interval $[a_1, a_2]$ is the union of $S_1, \dots, S_{n-1}$.

Suppose that, for some k , S_k contains a subinterval of $[a_1, a_2]$. Let $h + 1$ be the least value of k with this property and let $[b_1, b_2]$ be a subinterval contained in S_{h+1} . Thus

$$W_{h+1}(t, b) \equiv \begin{vmatrix} y_1(t, b) & \dots & y_{h+1}(t, b) \\ & \dots\dots & \\ y_1^{(h)}(t, b) & \dots & y_{h+1}^{(h)}(t, b) \end{vmatrix} = 0 \quad \text{for } b_1 \le t \le b_2 .$$

By the minimal nature of $h + 1$ there exists an interval $[c_1, c_2] \subset [b_1, b_2]$ on which

$$W_h(t, b) = \begin{vmatrix} y_1(t, b) & \dots & y_h(t, b) \\ & \dots\dots & \\ y_1^{(h-1)}(t, b) & \dots & y_h^{(h-1)}(t, b) \end{vmatrix} \neq 0 .$$

Consider the differential equation

$$\begin{vmatrix} y_1(t, b) & \dots & y_h(t, b) & y \\ & \dots\dots & & \\ y_1^{(h)}(t, b) & \dots & y_h^{(h)}(t, b) & y^{(h)} \end{vmatrix} = 0$$

on the interval $[c_1, c_2]$. It is of order h and has the linearly independent solutions $y_1, \dots, y_h$. It also has the solution y_{h+1} . Thus y_{h+1} is a linear combination of $y_1, \dots, y_h$ on $[c_1, c_2]$, which is a contradiction because $y_1, \dots, y_n$ are a fundamental system of solutions of (1).

Therefore none of the closed sets S_k contains a subinterval of $[a_1, a_2]$. If t_1 is a point not in S_1 there is a neighbourhood N_1 of t_1 not containing any point of S_1 . In N_1 there is a point t_2 not in S_2 and a neighbourhood N_2 of t_2 not containing any point of S_2 . Continuing in this way we finally obtain a neighbourhood N_{n-1} of a point t_{n-1} which does not contain any point of $S_1, \dots, S_{n-1}$. This is the final contradiction.

In the same way we can define first left conjugate points. For some b in I there exists $a < b$ in I such that (1) is not disconjugate on $[a, b]$. Moreover if b has this property, so does any $b' > b$ in I . We denote by $\eta_-(b)$ the infimum of all $c < b$ such that (1) is disconjugate on $[c, b]$. Thus if $\eta_-(b)$ is defined, so is $\eta_-(b')$ for any $b' > b$ in I and $\eta_-(b') \ge \eta_-(b)$. As before we can show that $\eta_-(b)$ is an increasing function.

LEMMA 13. *If* $b = \eta_+(a)$ *then* $a = \eta_-(b)$.

Proof. By Proposition 10 and Lemma 12 there exists a solution with the property

(k, a, b) for some k such that $1 \le k < n$. Since this solution also has the property $(n-k, b, a)$ it follows that $c = \eta_-(b) \ge a$. Similarly there exists a solution with the property $(n-h, b, c) = (h, c, b)$, and hence $\eta_+(c) \le b$. If we had $c > a$ we would have $\eta_+(c) > \eta_+(a) = b$. Therefore $c = a$.

In exactly the same way as we proved Theorem 7 of Chapter 2 we can now prove

THEOREM 7. *The function* $\eta(a)$ *is continuous and its domain is an open subinterval of* I.

As an application we prove

THEOREM 8. *The equation* (1) *is disconjugate on a half-open interval* $[a, b)$ *if, and only if, it is disconjugate on its interior* (a, b).

Proof. Extend the domain of definition of the coefficients of (1) by setting $p_k(t) = p_k(a)$ for $t < a$. Then the preceding results can be applied to the open interval $I = (-\infty, b)$. If the equation (1) is disconjugate on (a, b) but not on $[a, b)$ then $\eta(a)$ is defined but not $\eta(c)$ for any $c > a$. However this contradicts Theorem 7.

In a similar manner one can prove the analogue of Proposition 3 for half-open intervals.

PROPOSITION 12. *If every nontrivial solution has less than* n *distinct zeros on the half-open interval* $[a, b)$ *then every nontrivial solution has less than* n *zeros, counting multiplicities.*

5. Adjoint equations

We are going to consider the relation between the disconjugacy properties of an equation

$$Lu \equiv u^{(n)} + p_1(t)u^{(n-1)} + \dots + p_n(t)u = 0 \tag{10}$$

and those of its adjoint

$$L^*v \equiv (-1)^n v^{(n)} + (-1)^{n-1}[p_1(t)v]^{(n-1)} + \dots + p_n(t)v = 0 . \tag{11}$$

The equation (10) can be replaced in the usual way by an equivalent system

$$y' = A(t)y , \tag{12}$$

where

$$y = \begin{pmatrix} y_1 \\ y_2 \\ \vdots \\ y_{n-1} \\ y_n \end{pmatrix}, \quad A(t) = \begin{pmatrix} 0 & 1 & 0 & \dots & 0 \\ 0 & 0 & 1 & \dots & 0 \\ & & \dots\dots & & \\ 0 & 0 & 0 & \dots & 1 \\ -p_n & -p_{n-1} & -p_{n-2} & \dots & -p_1 \end{pmatrix},$$

and $y_k = u^{(k-1)}$. The adjoint equation (11) is equivalent to the system

$$z' = B(t)z, \tag{13}$$

where

$$z = \begin{pmatrix} z_1 \\ z_2 \\ \vdots \\ z_{n-1} \\ z_n \end{pmatrix}, \quad B(t) = \begin{pmatrix} p_1 & 1 & 0 & \dots & 0 \\ -p_2 & 0 & 1 & \dots & 0 \\ & & \dots\dots & & \\ (-1)^{n-2}p_{n-1} & 0 & 0 & \dots & 1 \\ (-1)^{n-1}p_n & 0 & 0 & \dots & 0 \end{pmatrix},$$

and $z_k = v^{(k-1)} - (p_1 v)^{(k-2)} + (p_2 v)^{(k-3)} - \dots + (-1)^{k-1}p_{k-1}v$. A solution of (10), (11) has a zero of multiplicity $\geq k$ at some point if and only if the first k coordinates of the corresponding solution of the system (12), (13) vanish at this point. If we put

$$K = K_n = \begin{pmatrix} 0 & 0 & \dots & 0 & -1 \\ 0 & 0 & \dots & 1 & 0 \\ & & \dots\dots & & \\ (-1)^n & 0 & \dots & 0 & 0 \end{pmatrix},$$

then it is easily verified that

$$B(t) = (-1)^n K A^*(t) K .$$

Moreover

$$K^2 = (-1)^{n-1} I .$$

Let $Y(t, a)$ be the fundamental matrix of (12) such that $Y(a, a) = I$. Then $Y^{*-1}(t, a)$ is the fundamental matrix of the equation $y' = -A^*(t)y$ which takes the value I at a and hence

$$Z(t, a) = (-1)^{n-1} K Y^{*-1}(t, a) K$$

is the fundamental matrix of (13) such that $Z(a, a) = I$. On the other hand

$Y(t, b) = Y(t, a)Y^{-1}(b, a)$ and hence $Y(a, b) = Y^{-1}(b, a)$. Consequently

$$Z(a, b) = (-1)^{n-1}KY^*(b, a)K . \tag{14}$$

If the equation (10) is not disconjugate on an interval I then for some k $(1 \leq k < n)$ and some a, b in I there exists a nontrivial solution with the property (k, a, b). Therefore the $k \times k$ submatrix Ω of $Y(b, a)$ formed by the first k rows and the last k columns has determinant zero. By (14) the corresponding submatrix of $Z(a, b)$ is $(-1)^{n-1}K_k\Omega^*K_k$, which also has determinant zero. Therefore the adjoint equation (11) has a nontrivial solution with the property (k, b, a) and is not disconjugate. Since $L^{**} = L$ we have proved

THEOREM 9. *The equation* (10) *is disconjugate on an interval* I *if and only if the adjoint equation* (11) *is disconjugate on* I. *If* (10) *has a nontrivial solution with the property* (k, a, b) *then* (11) *has a nontrivial solution with the property* (k, b, a).

We show next that if the operator L has a factorisation of the form (7) then its adjoint L^* has a factorisation of the same form in which the order of the factors is reversed.

THEOREM 10. *If*

$$Lu \equiv r_{n+1}\, D\, r_n\, D \dots D\, r_1 u ,$$

where $D = d/dt$, $r_k > 0$ *and* $r_k \in C^{\max(n-k+1,k-1)}$ *then*

$$L^*v \equiv (-1)^n\, r_1\, D\, r_2\, D \dots D\, r_{n+1}v .$$

Proof. If we put

$$L_0u = u , \quad L_ku = D\left(r_kL_{k-1}u\right) \quad (k = 1, \dots, n) ,$$

then $Lu = r_{n+1}L_nu$. Similarly, if we put

$$M_0v = r_{n+1}v , \quad M_kv = r_{n+1-k}D\left(M_{k-1}v\right) \quad (k = 1, \dots, n) ,$$

then

$$M_nv = Mv \equiv r_1\, D\, r_2\, D \dots D\, r_{n+1}v .$$

We have

$$D\left\{\sum_{k=0}^{n-1} (-1)^k M_k v . r_{n-k} L_{n-k-1} u\right\} = \sum_{k=0}^{n-1} (-1)^k \left\{D(M_k v) . r_{n-k} L_{n-k-1} u + M_k v . D(r_{n-k} L_{n-k-1} u)\right\}$$

$$= \sum_{k=0}^{n-1} (-1)^k \left\{M_{k+1} v . L_{n-k-1} u + M_k v . L_{n-k} u\right\}$$

$$= M_0 v . L_n u + (-1)^{n-1} M_n v . L_0 u$$

$$= v . Lu - (-1)^n Mv . u .$$

Therefore, by the uniqueness of the adjoint, $L^* = (-1)^n M$.

6. Green's functions

Throughout this section we suppose that *the equation* (1) *is disconjugate on the compact interval* $I = [a, b]$. Let $f(t)$ be a continuous function on I , let $a_1 < \ldots < a_m$ be points of I and let $r_1, \ldots, r_m$ be positive integers with sum n . The multipoint boundary value problem

$$Ly = f(t) , \tag{15}$$

$$y^{(\nu_i)}(a_i) = 0 \quad (\nu_i = 0, 1, \ldots, r_i-1;\ i = 1, \ldots, m) \tag{16}$$

has a unique solution $y(t)$, since the corresponding homogeneous problem has no nontrivial solution. This solution can be represented in the form

$$y(t) = \int_a^b G(t, s) f(s) ds , \tag{17}$$

where the *Green's function* $G(t, s)$ is defined by the properties

(i) as a function of t , $G(t, s)$ is a solution of the equation (1) on the intervals $[a, s)$ and $(s, b]$, satisfying the boundary conditions

$$G^{(\nu_i)}(a_i, s) = 0 \quad (\nu_i = 0, 1, \ldots, r_i-1;\ i = 1, \ldots, m) ;$$

(ii) as a function of t , $G(t, s)$ and its first $n - 2$ derivatives are continuous at $t = s$, while

$$G^{(n-1)}(s+0, s) - G^{(n-1)}(s-0, s) = 1 .$$

In fact it is readily verified by differentiating under the integral sign that if G has these properties then the function $y(t)$ defined by (17) is a solution of the boundary-value problem (15) - (16). It remains to show that there is a unique function G with the properties (i), (ii).

Let $y_1(t), \ldots, y_n(t)$ be a fundamental system of solutions of the homogeneous equation (1), e.g., the system which is determined by the initial conditions

$$y_k^{(i-1)}(a) = \delta_{ik} \quad (i, k = 1, \ldots, n) ,$$

and set

$$G(t, s) = \begin{cases} \alpha_1(s)y_1(t) + \ldots + \alpha_n(s)y_n(t) & \text{for} \quad t \leq s , \\ \beta_1(s)y_1(t) + \ldots + \beta_n(s)y_n(t) & \text{for} \quad t \geq s , \end{cases}$$

where the coefficients $\alpha_i(s)$, $\beta_i(s)$ are to be determined. If we put

$$\gamma_i(s) = \beta_i(s) - \alpha_i(s) ,$$

condition (ii) takes the form

$$\sum_{i=1}^{n} \gamma_i(s)y_i^{(j)}(s) = 0 \quad (j = 0, 1, \ldots, n-2)$$

$$\sum_{i=1}^{n} \gamma_i(s)y_i^{(n-1)}(s) = 1 .$$

The determinant of this system of linear equations is the Wronskian $W(y_1, \ldots, y_n)$. Therefore the differences $\gamma_i(s)$ are uniquely determined and are continuous functions of s. If we substitute $\beta_i(s) = \alpha_i(s) + \gamma_i(s)$ the boundary conditions in (i) provide n linear equations for the n unknowns $\alpha_i(s)$. These equations have a unique solution because the corresponding homogeneous equations have no nontrivial solution.

We are going to study now the zeros of the Green's function $G(t, s)$ for fixed s. Put

$$P(t) = (t-a_1)^{r_1} \ldots (t-a_m)^{r_m} . \tag{18}$$

LEMMA 14. *In the square* $a \leq t, s \leq b$

$$G(t, s)P(t) \geq 0 .$$

Proof. Let c be any point of I distinct from $a_1, \ldots, a_m$. If the function $G(c, s)$ were not of constant sign in I we could find a continuous function $f(t) > 0$ on I such that

$$\int_a^b G(c, s)f(s)ds = 0 .$$

Then the solution $y(t)$ of the boundary value problem (15) - (16) would have n zeros at the points a_i $(i = 1, \ldots, m)$ and a further zero at c. Therefore, by Proposition 7, the function $Ly = f$ has at least one zero in I, which is a contradiction. Thus $G(c, s)$ is of constant sign in I. To determine this sign connect the operator L to the operator D^n by a path $L^{(\lambda)}$, as in the proof of Proposition 9. If the boundary conditions (16) are kept fixed the corresponding Green's function $G^{(\lambda)}(t, s)$ is a continuous function of λ, by the manner in which it has been constructed. Hence $G(c, s) = G^{(1)}(c, s)$ has the same constant sign as $G^{(0)}(c, s)$. But the solution of the equation $D^n y = 1$ which satisfies the boundary conditions is $y(t) = P(t)/n!$. Since $y(t) = \int_a^b G^{(0)}(t, s)ds$ it follows that $G(c, s)$ has the same sign as $P(c)$.

LEMMA 15. *If* $a_1 < s < a_m$ *each zero of* $g(t) = G(t, s)$ *is isolated.*

Proof. Suppose on the contrary that there exists a sequence $\{t_\nu\}$ of points of I such that $t_\nu \to c$ and $g(t_\nu) = 0$. We will assume for definiteness that $c \geq s$ and $t_\nu > s$. Then $g(t)$ coincides on the interval $(s, b]$ with a solution of the equation (1) which has infinitely many zeros. Therefore $g(t) = 0$ for $t \geq s$. Since $g^{(k)}(t)$ $(k = 0, 1, \ldots, n-2)$ is continuous at $t = s$ it follows that $g(t)$ coincides on the interval $[a, s)$ with a solution of the equation (1) which has a zero of multiplicity $\geq n-1$ at s. Since $g(a_1) = 0$ also and the equation (1) is disconjugate on the interval $[a, s]$ this implies that $g(t) = 0$ for $t \leq s$. Therefore $g(t) \equiv 0$, which is impossible because $g^{(n-1)}(t)$ has a jump discontinuity at $t = s$.

Although $g(t)$ has only isolated zeros some functions related to it may vanish throughout an interval. It is convenient for our present purposes to regard such an interval as a single zero. More precisely, we define a *zero component* of a continuous function to be a maximal interval throughout which the function vanishes.

The operator L has a factorisation

$$Ly = \rho_{n+1} D \rho_n D \ldots D \rho_1 y ,$$

where $\rho_k > 0$ and $\rho_k \in C^{n-k+1}$ $(k = 1, \ldots, n)$, and $\rho_{n+1} = (\rho_1 \ldots \rho_n)^{-1}$. Put

$$L_0 y = y , \quad L_k y = D(\rho_k L_{k-1} y) \quad (k = 1, \ldots, n) ,$$

so that $Ly = \rho_{n+1} L_n y$.

LEMMA 16. *If* $a < s < b$ *and* $g(t) = G(t, s)$ *then* $L_{n-2}g(t)$ *has at most two zero components on the interval* $[a, b]$. *If it has two, then each is a simple zero.*

Proof. On each of the intervals $[a, s)$ and $(s, b]$, $g(t)$ is a solution of the equation $L_n y = 0$. Therefore $D(\rho_n L_{n-1} g) = 0$ and $\rho_n L_{n-1} g$ is a constant on each of the intervals $[a, s)$ and $(s, b]$. Hence on each of these intervals $D(\rho_{n-1} L_{n-2} g)$ has strict constant sign or is identically zero, and $\rho_{n-1} L_{n-2} g$ is strictly monotonic or constant. The result now follows at once.

THEOREM 11. $G(t, s)/P(t) > 0$ *for* $a_1 < s < a_m$, $a_1 \le t \le a_m$.

Proof. Put $J = [a_1, a_m]$. We show first that $G(t, s) \neq 0$ for $a_1 < t$, $a_1 < t, s < a_m$ and $t \neq a_2, \ldots, a_{m-1}$. Suppose on the contrary that for some s in (a_1, a_m) the function $g(t) = G(t, s)$ vanishes at a point of J distinct from $a_1, \ldots, a_m$. Then $g(t)$ has at least $n + 1$ zeros in J , including $m + 1$ distinct zeros. Therefore, by Rolle's theorem, $L_1 g$ has at least m zero components, each strictly interior to an interval between consecutive zeros of g . Let m_1 be the number of distinct points a_k which are zeros of g of multiplicity ≥ 2 . All such points are zeros of $L_1 g$ with total multiplicity $n - m$. Thus $L_1 g$ has at least $m + m_1$ zero components and at least n zeros. By Rolle's theorem, $L_2 g$ has at least $m + m_1 - 1$ zero components, besides those points a_k which are zeros of g of multiplicity ≥ 3 . Let m_2 be the number of such points. They are all zeros of $L_2 g$ with total multiplicity $n - m - m_1$. Thus $L_2 g$ has at least $m + m_1 + m_2 - 1$ zero components and at least $n - 1$ zeros. Continuing in this way we see that $L_{n-2} g$ has at least $m + m_1 + \ldots + m_{n-2} - (n-3)$ zero components. But $m + m_1 + \ldots + m_{n-2} = n$, since in the sum on the left each point a_k is counted r_k times. Thus $L_{n-2} g$ has at least 3 zero components, which contradicts Lemma 16.

Similarly we can show that $G^{(r_k)}(a_k, s) \neq 0$. If for some s in (a_1, a_m) and some k $(1 \le k \le m)$ this were not so then the isolated zeros a_j of $g(t)$ have total multiplicity at least $n + 1$. By successive applications of Rolle's theorem we find that $L_{n-2} g$ has at least 2 zero components and at least 3 zeros, which contradicts Lemma 16.

Thus $G(t, s)$ and $P(t)$ have the same zeros with the same multiplicities.

Consequently the ratio $G(t, s)/P(t)$ is finite and of strict constant sign. Hence, by Lemma 14, it is positive.

The following result on differential inequalities is a simple application of Theorem 11.

PROPOSITION 13. *Suppose the equation* (1) *is disconjugate on the interval* I. *Suppose further that* $y(t)$ *does not vanish throughout any subinterval of* I, *that it has at least two distinct zeros, and that* $Ly \geq 0$ *throughout* I. *Then* $y(t)$ *has at most* n *zeros on* I. *If* $y(t)$ *has zeros* $a_1 < \ldots < a_m$ *with multiplicities* $r_1, \ldots, r_m$, *where* $r_1 + \ldots + r_m = n$ *and* $m > 1$, *then*

$$y(t)/P(t) > 0 \quad \textit{throughout } I. \tag{19}$$

Proof. Put $f = Ly$ and suppose $y(t)$ has zeros $a_1 < \ldots < a_m$ with multiplicities $\geq r_1, \ldots, r_m$ respectively, where $r_1 + \ldots + r_m = n$. Then for $a_1 \leq t \leq a_m$,

$$y(t) = \int_{a_1}^{a_m} G(t, s)f(s)ds .$$

If $f(t) \equiv 0$ on the subinterval $J = [a_1, a_m]$ then $y(t)$ is a solution of the equation (1) with n zeros on J. Therefore $y(t) \equiv 0$ on J, which is contrary to hypothesis. Hence $f(s) > 0$ for some s in J. It follows from Theorem 11 that (19) holds for $a_1 \leq t \leq a_m$. Thus the zeros a_k have multiplicities exactly r_k $(k = 1, \ldots, m)$ and there are no further zeros on I. Since $P(t)$ also has no further zeros the inequality (19) continues to hold throughout I.

7. Principal solutions

Let I be an interval open at its right endpoint b and let a be a fixed point of I. For any $c > a$ the equation (1) has a unique solution $u(t, c)$ which satisfies the initial conditions

$$u = u' = \ldots = u^{(n-2)} = 0 , \quad (-1)^{n-1}u^{(n-1)} > 0 \quad \text{at} \quad t = c ,$$

with the normalisation

$$u^2 + u'^2 + \ldots + u^{(n-1)^2} = 1 \quad \text{at} \quad t = a .$$

Now give c a sequence of values converging to b. By restricting attention to a suitable subsequence $\{c_\nu\}$ we can ensure that

$$\eta_j = \lim_{\nu\to\infty} u^{(j-1)}(a, c_\nu) \quad (j = 1, \ldots, n)$$

exists. Then $\eta_1^2 + \ldots + \eta_n^2 = 1$ and, if $U(t)$ is the solution of (1) such that $U^{(j-1)}(a) = \eta_j$, we have $u^{(j-1)}(t, c_\nu) \to U^{(j-1)}(t)$ uniformly on compact subintervals of I as $\nu \to \infty$ $(j = 1, \ldots, n)$. We are going to show that if (1) has a Markov fundamental system of solutions on I then it is unnecessary to restrict attention to a subsequence $\{c_\nu\}$.

THEOREM 12. *Let I be an interval open at its right endpoint b and let a be a fixed point of I. Suppose the equation* (1) *has a Markov fundamental system of solutions on I. If $u(t, c)$ is the uniquely determined solution of* (1) *such that*

$$u = u' = \ldots = u^{(n-2)} = 0, \quad (-1)^{n-1}u^{(n-1)} > 0 \quad at \quad t = c,$$

$$u^2 + u'^2 + \ldots + u^{(n-1)^2} = 1 \qquad at \quad t = a,$$

then

$$U^{(j-1)}(t) = \lim_{c\to b} u^{(j-1)}(t, c) \quad (j = 1, \ldots, n)$$

exists uniformly on compact subintervals of I. The solution $U(t)$ of (1) *has the following properties:*

(i) $U > 0$ *throughout* I,

(ii) if $u_1, u_2, \ldots, u_n$ is a Markov fundamental system of solutions of (1) *then*

$$W(U, u_1, \ldots, u_k) \geq 0 \quad (k = 1, \ldots, n-1) ;$$

moreover there exists a Markov fundamental system of solutions $u_1, u_2, \ldots, u_n$ with $u_1 = U$,

(iii) if $u_1, u_2, \ldots, u_n$ is a Descartes fundamental system of solutions of (1) *then $U, u_2, \ldots, u_n$ is also a Descartes fundamental system of solutions.*

Proof. Let $u_1, \ldots, u_n$ be any Markov fundamental system of solutions, so that

$$W_k = W(u_1, \ldots, u_k) > 0 \quad (k = 1, \ldots, n) .$$

By Theorem 2 the linear differential operator L has the representation

$$Lu \equiv v_1 \ldots v_n D \frac{1}{v_n} D \ldots D \frac{1}{v_1} u ,$$

where $v_1 = W_1 > 0$, $v_k = W_k W_{k-2}/W_{k-1}^2 > 0$ $(k = 2, \ldots, n)$. Moreover, by the proof of Theorem 2, if we put

$$y_1 = u/u_1 , \quad y_k = W(u, u_1, \ldots, u_{k-1})/W_k \quad (k = 2, \ldots, n) ,$$

then

$$y_k = (-1)^{k-1} \frac{1}{v_k} D \frac{1}{v_{k-1}} D \ldots D \frac{1}{v_1} u .$$

Hence the equation (1) is equivalent to the system

$$\begin{aligned} y_k' &= - v_{k+1} y_{k+1} \quad (k = 1, \ldots, n-1) , \\ y_n' &= 0 . \end{aligned} \tag{20}$$

The solution $y(t, c)$ of the system (20) corresponding to the solution $u(t, c)$ of (1) has the coordinates

$$y_1 = y_2 = \ldots = y_{n-1} = 0 , \quad y_n = (-1)^{n-1} u^{(n-1)} W_{n-1} > 0 \quad \text{at} \quad t = c .$$

By changing t into $-t$ in (20) we obtain a system of type K , in the terminology of Coppel [1]. For two vectors y, z we write $y \geq z$ if the coordinates of the vector $y - z$ are all non-negative. Since $y(t, c) \geq 0$ for $t = c$ it follows from the basic property of systems of type K that $y(t, c) \geq 0$ for $t \leq c$. Hence for the solution $Y(t)$ of (20) corresponding to the solution $U(t)$ of (1) we have $Y(t) \geq 0$ for all t in I . Moreover, if $Y_1(s) = 0$ then, since $Y_1 \geq 0$ and $Y_1' \leq 0$, we must have $Y_1(t) = 0$ for $s \leq t < b$. Therefore $U(t) \equiv 0$, which is a contradiction. Thus, throughout I ,

$$U > 0 , \quad W(U, u_1, \ldots, u_k) \geq 0 \quad (k = 1, \ldots, n-1) . \tag{21}$$

In particular, for $k = 1$,

$$U'/U \leq u_1'/u_1 .$$

Similarly, if $Y_k(s) = 0$ we must have $Y_k(t) = 0$ for $s \leq t < b$. Since U is a solution of (1) and $u_1, \ldots, u_n$ is a fundamental system of solutions there is a uniquely determined integer h $(1 \leq h \leq n)$ such that

$$U = \alpha_1 u_1 + \ldots + \alpha_h u_h , \quad \alpha_h \neq 0 .$$

Since $u_1, \ldots, u_n$ form a Markov system it follows that, throughout I ,

$$W(U, u_1, \ldots, u_k) > 0 \quad \text{for} \quad 1 \leq k < h$$

and

$$W(U, u_1, \ldots, \hat{u}_h, \ldots, u_k) = (-1)^{h-1}\alpha_h W_k \neq 0 \quad \text{for} \quad h < k \leq n ,$$

where $\hat{u}_h$ denotes that u_h is omitted. Thus there is a Markov fundamental system of solutions $U, u_1, \ldots, u_{h-1}, \pm u_{h+1}, \ldots, \pm u_n$ with first term U.

Hence $U'/U = \min(u_1'/u_1)$, where the minimum is taken over all Markov fundamental systems $u_1, \ldots, u_n$. This characterizes the solution U independently of the sequence $\{c_\nu\}$ which was used to construct it. It follows that if U_1, U_2 are the solutions corresponding to two sequences $\{c_\nu'\}, \{c_\nu''\}$ then $U_1'/U_1 = U_2'/U_2$, and hence U_1/U_2 is a positive constant. This constant must be 1 , because of the normalisation at $t = a$. Thus the limit U is independent of the sequence $\{c_\nu\}$.

Suppose finally that $u_1, \ldots, u_n$ is a Descartes fundamental system of solutions. By Theorem 4 such a system certainly exists if I is half-open. To prove that $U, u_2, \ldots, u_n$ is a Descartes fundamental system of solutions it is sufficient, by Proposition 4, to show that

$$W(U, u_2, \ldots, u_k) > 0 \quad (k = 2, \ldots, n) . \tag{22}$$

We prove this by induction on k . By Lemma 5 we have

$$W(u_1, \ldots, u_{k-1})W(U, u_2, \ldots, u_k) = W(U, u_2, \ldots, u_{k-1})W(u_1, \ldots, u_k) + W(u_2, \ldots, u_k)W(U, u_1, \ldots, u_{k-1}) .$$

Since $W(U, u_1, \ldots, u_{k-1}) \geq 0$, by (21), it follows that (22) holds for $k = 2$ and that it holds for k if it holds for $k - 1$.

The uniquely determined solution U will be called the *principal solution* of (1). We will now show that it is the 'smallest' solution in the neighbourhood of b .

THEOREM 13. *Let I be an interval open at its right endpoint b and let a be a fixed point of I . Suppose the equation (1) has a Markov fundamental system of solutions on I . Then it has a Markov fundamental system of solutions $U_1, \ldots, U_n$ satisfying the additional conditions:*

[A] *U_k has a zero of multiplicity $k - 1$ at a $(k = 1, \ldots, n)$,*

[B] *$U_k > 0$ near b $(k = 1, \ldots, n)$ and*

$$U_{k-1} = o(U_k) \quad \text{for} \quad t \to b \quad (k = 2, \ldots, n) .$$

Moreover $U_1, \ldots, U_n$ is a Descartes system for $a < t < b$.

Proof. Since the result is trivial for $n = 1$, we assume that $n > 1$ and that it holds for equations of order less than n. Let $U_1(t)$ be the principal solution of (1) and put $v = (u/U_1)'$. Then $Lu \equiv Mv$, where M is a linear differential operator of order $n - 1$ with continuous coefficients and leading coefficient 1. By Theorem 12 the equation (1) has a Markov fundamental system of solutions $u_1, u_2, \ldots, u_n$ with $u_1 = U_1$. If we put $v_k = (u_{k+1}/U_1)'$ $(k = 1, \ldots, n-1)$ then, by Lemma 3, the equation

$$Mv = 0 \tag{23}$$

has the Markov fundamental system of solutions $v_1, \ldots, v_{n-1}$. Therefore, by the induction hypothesis, (23) has a Markov fundamental system of solutions $V_1, \ldots, V_{n-1}$ such that V_k has a zero of multiplicity $k - 1$ at a and $V_k > 0$ for $a < t < b$ $(k = 1, \ldots, n-1)$, and

$$V_{k-1} = o(V_k) \quad \text{for} \quad t \to b \quad (k = 2, \ldots, n-1) .$$

Put

$$U_k(t) = U_1(t) \int_a^t V_{k-1}(s)ds \quad (k = 2, \ldots, n) .$$

Then $U_1, \ldots, U_n$ is a Markov fundamental system of solutions of (1), U_k has a zero of multiplicity $k - 1$ at a and $U_k > 0$ for $a < t < b$ $(k = 1, \ldots, n)$. Hence, by Lemma 8, $U_1, \ldots, U_n$ is a Descartes system on (a, b). Finally

$$U_2(t)/U_1(t) = \int_a^t V_1(s)ds$$

is an increasing function for $t > a$. Suppose

$$\gamma = \int_a^b V_1(s)ds < \infty .$$

Then $\gamma U_1 > U_2$ throughout (a, b) and $\gamma U_1 - U_2$, $U_2, \ldots, U_n$ is a Markov fundamental system of solutions of (1) on this interval. Since the principal solution is uniquely determined it follows from Theorem 12, with $u_1 = \gamma U_1 - U_2$, that

$$0 \le W(U_1, \gamma U_1 - U_2) = - W(U_1, U_2) ,$$

which is a contradiction. We conclude that $\gamma = + \infty$ and $U_1 = o(U_2)$ for $t \to b$. If $1 < k < n$ then, by the theory of indeterminate forms,

$$U_k(t)/U_{k+1}(t) = \int_a^t V_{k-1}(s)ds \Big/ \int_a^t V_k(s)ds \to 0 \quad \text{as} \quad t \to b \ .$$

This completes the proof.

Evidently any set of solutions $U_1, \ldots, U_n$ satisfying either [A] or [B] is a fundamental system. The condition [B] uniquely determines U_1 up to a positive constant factor and thus gives another characterization of the principal solution. It determines U_k $(k > 1)$ up to a positive constant factor and addition of a linear combination of $U_1, \ldots, U_{k-1}$. The conditions [A] and [B] together determine all U_k up to positive constant factors. It should be observed that [A] says almost the same thing relative to the point a as [B] relative to the point b.

As an application of Theorem 13 we prove

PROPOSITION 14. *Let I be an interval open at its right endpoint b and let $u_1, \ldots, u_n$ be C^n-functions which are positive near b and satisfy $u_{k-1} = o(u_k)$ for $t \to b$ $(k = 2, \ldots, n)$.*

If $W(u_1, \ldots, u_k) \neq 0$ on I $(k = 1, \ldots, n)$ then $u_1, \ldots, u_n$ is a Markov system on I and a Descartes system near b.

If $W(u_k, \ldots, u_n) \neq 0$ on I $(k = 1, \ldots, n)$ then $u_1, \ldots, u_n$ is a Descartes system on I.

Proof. We use induction on n, the result being easily verified for $n = 2$. The equation $W(u_1, \ldots, u_n, u) = 0$ has a Markov fundamental system of solutions on I, $\pm u_1, \ldots, \pm u_n$ if $W(u_1, \ldots, u_k) \neq 0$ $(k = 1, \ldots, n)$ and $\pm u_n, \ldots, \pm u_1$ if $W(u_k, \ldots, u_n) \neq 0$ $(k = 1, \ldots, n)$. Therefore, by Theorem 13 it has a Markov fundamental system of solutions $U_1, \ldots, U_n$ satisfying [B]. It follows that

$$u_k = \alpha_k U_k + \text{linear combination of } U_1, \ldots, U_{k-1} ,$$

where $\alpha_k > 0$ $(k = 1, \ldots, n)$. Hence $u_1, \ldots, u_n$ also form a Markov system. Therefore, by the induction hypothesis, $u_1, \ldots, u_{n-1}$ form a Descartes system near b. By Lemma 4

$$\left[\frac{W(u_k,\ldots,u_n)}{W(u_{k-1},\ldots,u_{n-1})}\right]' = \frac{W(u_k,\ldots,u_{n-1})W(u_{k-1},\ldots,u_n)}{[W(u_{k-1},\ldots,u_{n-1})]^2} , \tag{24}$$

which shows that $W(u_k, \ldots, u_n) \neq 0$ near b if $W(u_{k-1}, \ldots, u_n) \neq 0$ near b. Since $W(u_1, \ldots, u_n) > 0$ it follows that $W(u_k, \ldots, u_n) \neq 0$ near b for

$k = 1, \ldots, n$. Therefore, by the induction hypothesis, $u_2, \ldots, u_n$ is a Descartes system near b . From Proposition 4 we conclude that $u_1, \ldots, u_n$ is a Descartes system near b .

Suppose now that $W(u_k, \ldots, u_n) \neq 0$ on I $(k = 1, \ldots, n)$. Then, by the induction hypothesis, $u_2, \ldots, u_n$ is a Descartes system on I . To prove that $u_1, \ldots, u_n$ is a Descartes system on I it is sufficient to show that the inequalities $W(u_1, \ldots, u_k) > 0$ $(k = 1, \ldots, n)$ hold not only near b but throughout I . This is certainly true for $k = n$, since $W(u_1, \ldots, u_n) \neq 0$ on I . Moreover it is true for k if it is true for $k + 1$, since by Lemma 4

$$\left[\frac{W(u_1,\ldots,u_k)}{W(u_2,\ldots,u_{k+1})}\right]' = -\frac{W(u_2,\ldots,u_k)W(u_1,\ldots,u_{k+1})}{[W(u_2,\ldots,u_{k+1})]^2} . \tag{25}$$

Hence it holds for all k , and the proof is complete.

We now use this result to study the properties of solutions of (1) which satisfy [B] but not necessarily [A].

THEOREM 14. *Let I be an interval open at its right endpoint b and suppose the equation* (1) *has a Markov fundamental system of solutions on I . Then any set of solutions satisfying* [B] *is a Markov fundamental system on I and a Descartes system near b .*

Let $U_1, \ldots, U_n$ be solutions satisfying [B] *and forming a Descartes system on (a, b) . If $i_1 < \ldots < i_k$ and $j_1 < \ldots < j_k$ are distinct k-tuples such that $i_\alpha \leq j_\alpha$ $(\alpha = 1, \ldots, k)$ then the ratio*

$$W\left(U_{i_1}, \ldots, U_{i_k}\right)\Big/W\left(U_{j_1}, \ldots, U_{j_k}\right)$$

is a decreasing function on (a, b) and tends to zero as $t \to b$.

Proof. By Theorem 13 the equation (1) has a Markov fundamental system of solutions on I which satisfies [B]. Since [B] uniquely determines U_k up to a positive constant factor and addition of a linear combination of $U_1, \ldots, U_{k-1}$, it follows that any set of solutions satisfying [B] is a Markov system on I . Therefore, by Proposition 14, it is a Descartes system near b .

To prove the second statement it is sufficient to consider the case where $i_\beta < j_\beta$ for some β and $i_\alpha = j_\alpha$ for all $\alpha \neq \beta$, since by replacing i_k by j_k , then i_{k-1} by j_{k-1} and so on, the general case can be reduced to this special case. By Lemma 4

$$\left[\frac{W\left(U_{i_1},\ldots,U_{i_{\beta-1}},U_{j_\beta},U_{i_{\beta+1}},\ldots,U_{i_k}\right)}{W\left(U_{i_1},\ldots,U_{i_k}\right)}\right]' =$$

$$\frac{W\left(U_{i_1},\ldots,U_{i_{\beta-1}},U_{i_{\beta+1}},\ldots,U_{i_k}\right)W\left(U_{i_1},\ldots,U_{i_{\beta-1}},U_{i_\beta},U_{j_\beta},U_{i_{\beta+1}},\ldots,U_{i_k}\right)}{\left[W\left(U_{i_1},\ldots,U_{i_k}\right)\right]^2} .$$

It follows that the ratio $W\left(U_{j_1}, \ldots, U_{j_k}\right)\Big/W\left(U_{i_1}, \ldots, U_{i_k}\right)$ is an increasing function and tends to a limit γ $(0 < \gamma \leq \infty)$ as $t \to b$. If $\gamma < \infty$ then for $a < t < b$

$$W\left(U_{j_1}, \ldots, U_{j_k}\right) < \gamma W\left(U_{i_1}, \ldots, U_{i_k}\right) ,$$

that is

$$\tilde{W} \equiv W\left(U_{i_1}, \ldots, U_{i_{\beta-1}}, U_{j_\beta}-\gamma U_{i_\beta}, U_{i_{\beta+1}}, \ldots, U_{i_k}\right) < 0 .$$

Since $U_{i_\beta}/U_{j_\beta} \to 0$ as $t \to b$ there is a neighbourhood of b on which $U_{j_\beta} - \gamma U_{i_\beta} > 0$. In this neighbourhood the functions $U_{i_1}, \ldots, U_{i_{\beta-1}}$, $U_{j_\beta}-\gamma U_{i_\beta}, U_{i_{\beta+1}}, \ldots, U_{i_k}$ are solutions satisfying [B] of an equation of order k which has the Markov fundamental system of solutions $U_1, \ldots, U_{i_{\beta-1}}, U_{i_{\beta+1}}, \ldots$, $U_{i_k}, \pm\left(U_{j_\beta}-\gamma U_{i_\beta}\right)$. Applying the first part of the theorem to this equation we obtain $\tilde{W} > 0$, which is a contradiction. Therefore $\gamma = \infty$.

As a further application of Theorem 13 we have

THEOREM 15. *The equation* (1) *has a Markov fundamental system of solutions on an open interval* I *if, and only if, it is disconjugate on* I .

Proof. Let b be the right endpoint of I and let a be any point of I . If $c < a$ and $c \in I$ then by Lemma 6 the equation (1) has a Markov fundamental system of solutions on the open interval (c, b) and hence on the interval $[a, b)$. Therefore, by Theorem 13, it has a Markov fundamental system of solutions $U_1, \ldots, U_n$ on $[a, b)$ which satisfies [B]. Since this holds for each a in I and since [B] uniquely determines U_k up to a positive constant factor and addition of a linear combination of $U_1, \ldots, U_{k-1}$ it follows that $U_1, \ldots, U_n$ form a

Markov system throughout I .

The existence of a Markov fundamental system of solutions does not imply the existence of a Descartes fundamental system of solutions on an open interval, as it does for an interval which is not open (Theorem 4). For example, the equation $y'' = 0$ is disconjugate on the open interval $I = (-\infty, \infty)$ and has the Markov fundamental system of solutions $y_1 = 1$, $y_2 = t$. But it is clear that it has no Descartes fundamental system of solutions on I . However, if we assume the existence of a Descartes fundamental system of solutions we can obtain an extension of Theorem 13.

PROPOSITION 15. *Suppose the equation* (1) *is disconjugate on the open interval* $I = (a, b)$. *Then any set of solutions* $U_1, \ldots, U_n$ *which are positive near* a *and* b *and satisfy*

$$\left.\begin{aligned} U_k &= o(U_{k-1}) \quad \text{for} \quad t \to a , \\ U_{k-1} &= o(U_k) \quad \text{for} \quad t \to b , \end{aligned}\right\} \qquad (k = 2, \ldots, n)$$

is a Descartes fundamental system of solutions on I .

Proof. The equation (1) has a Markov fundamental system of solutions on the open interval I . Therefore, by Theorem 14, $U_1, \ldots, U_n$ is a Markov fundamental system of solutions on I . If we put

$$V_k(t) = U_{n-k+1}(-t) ,$$

then $V_1, \ldots, V_n$ satisfy the second condition of Proposition 14 on the interval $(-b, -a)$, since

$$W(V_k, \ldots, V_n) = W(U_1, \ldots, U_{n-k+1}) > 0 \quad (k = 1, \ldots, n) .$$

Consequently $V_1, \ldots, V_n$ is a Descartes system on the interval $(-b, -a)$, and hence $U_1, \ldots, U_n$ is a Descartes system on the interval (a, b) .

THEOREM 16. *Suppose the equation* (1) *has a Descartes fundamental system of solutions on the open interval* $I = (a, b)$. *Then it has a Descartes fundamental system of solutions* $U_1, \ldots, U_n$ *satisfying the additional conditions:*

$$\left.\begin{aligned} U_k &= o(U_{k-1}) \quad \text{for} \quad t \to a , \\ U_{k-1} &= o(U_k) \quad \text{for} \quad t \to b . \end{aligned}\right\} \qquad (k = 2, \ldots, n)$$

Proof. We use induction on the order n . As in the proof of Theorem 13 let U_1 be the principal solution of (1) and put $v = (u/U_1)'$, so that $Lu \equiv Mv$. By Theorem 12 the equation (1) has a Descartes fundamental system of solutions

$u_1, \ldots, u_n$ with $u_1 = U_1$. Then $v_1, \ldots, v_{n-1}$, where $v_k = (u_{k+1}/U_1)'$, is a Descartes fundamental system of solutions of the equation (23). Therefore, by the induction hypothesis, (23) has a Descartes fundamental system of solutions $V_1, \ldots, V_{n-1}$ such that

$$\begin{aligned} V_k &= o(V_{k-1}) \quad \text{for} \quad t \to a , \\ V_{k-1} &= o(V_k) \quad \text{for} \quad t \to b . \end{aligned} \qquad (k = 2, \ldots, n-1)$$

Since

$$W(U_1, u_k) = U_1 u_k' - U_1' u_k > 0$$

u_k/U_1 decreases with t and has a finite non-negative limit γ_k as $t \to a$ $(k = 2, \ldots, n)$. Hence for any solution $u = \alpha_1 u_1 + \ldots + \alpha_n u_n$ of (1) the ratio u/U_1 has a finite limit, namely, $\alpha_1 + \alpha_2\gamma_2 + \ldots + \alpha_n\gamma_n$, as $t \to a$. Thus for any solution v of (23), $\int_a^t v(s)ds$ exists as a finite limit.

Put

$$U_k(t) = U_1(t) \int_a^t V_{k-1}(s)ds \quad (k = 2, \ldots, n) .$$

Then $U_1, \ldots, U_n$ are a Markov fundamental system of positive solutions of (1), and $U_2 = o(U_1)$ for $t \to a$. If $1 < k < n$ then, by the theory of indeterminate forms,

$$U_{k+1}(t)/U_k(t) = \int_a^t V_k(s)ds \Big/ \int_a^t V_{k-1}(s)ds \to 0 \quad \text{as} \quad t \to a .$$

Exactly as in the proof of Theorem 13 we can show that

$$U_k = o(U_{k+1}) \quad \text{for} \quad t \to b \quad (k = 1, \ldots, n-1) .$$

Hence, by Proposition 15, $U_1, \ldots, U_n$ is a Descartes system on I.

If a second order equation is non-oscillatory on a half-open interval $[a, b)$ then it is disconjugate on a subinterval $[c, b)$. The adjoint equation is then disconjugate on $[c, b)$ and hence non-oscillatory on $[a, b)$. We give finally an example which shows that these properties cannot be extended to higher order equations.

The functions

$$u_1 = 1 , \quad u_2 = t^{1/2} + \sin t , \quad u_3 = t - 2\cos t$$

have Wronskian

$$
\begin{aligned}
W(u_1, u_2, u_3) &= u_2'u_3'' - u_2''u_3' \\
&= 2 + \sin t + t^{-1/2}\cos t + t^{-3/2}(1+2\sin t)/4 \\
&> 0 \quad \text{for} \quad \pi \le t < \infty .
\end{aligned}
$$

Hence u_1, u_2, u_3 are a fundamental system of solutions on the interval $[\pi, \infty)$ of a third order equation $Lu = 0$ with continuous coefficients and leading coefficient 1 . Every nontrivial solution has at most finitely many zeros on $[\pi, \infty)$, since if it is not constant it behaves asymptotically like $\alpha t^{1/2}$ or αt $(\alpha \neq 0)$. However, the equation $Lu = 0$ is not disconjugate on any subinterval (c, ∞) . For if it were, it would possess a Markov fundamental system of solutions on this subinterval and Theorem 14 would imply that $W(u_1, u_2) > 0$ for all large t . But

$$
W(u_1, u_2) = u_2' = t^{-1/2}/2 + \cos t ,
$$

which changes sign for arbitrarily large t .

The adjoint equation $L^*v = 0$ has the fundamental system of solutions

$$
\begin{aligned}
v_1 &= W(u_3, u_2) = t[\cos t + o(1)] \\
v_2 &= W(u_1, u_3) = 1 + 2\sin t \\
v_3 &= W(u_2, u_1) = -\cos t + o(1) .
\end{aligned}
$$

Thus every solution has the form

$$
\begin{aligned}
v &= \alpha_1 v_1 + \alpha_2 v_2 + \alpha_3 v_3 \\
&= \alpha_1 t[\cos t + o(1)] + \alpha_2[1+2\sin t] + \alpha_3[-\cos t + o(1)]
\end{aligned}
$$

and hence has infinitely many zeros.

8. Differential inequalities

For second order equations disconjugacy is implied by the existence of a solution of the Riccati inequality $R[w] \le 0$ or, equivalently, by the existence of a function $u > 0$ such that $Lu \le 0$. In this section we consider the extension of this result to equations of arbitrary order.

Let $u_1, \ldots, u_n$ be C^n-functions such that throughout the interval I

$$
W_k = W(u_1, \ldots, u_k) > 0 \quad (k = 1, \ldots, n) ,
$$

i.e., $u_1, \ldots, u_n$ is a Markov system, and let

$$Lu \equiv u^{(n)} + p_1(t)u^{(n-1)} + \ldots + p_n(t)u$$

be an n-th order linear differential operator with continuous coefficients p_k $(k = 1, \ldots, n)$. Since

$$W(u, u_1, \ldots, u_k) = (-1)^k W_k u^{(k)} + \text{linear combination of } u, u', \ldots, u^{(k-1)}$$

we can write

$$Lu \equiv c_0(t)u/u_1 + \sum_{k=1}^{n-1} c_k(t)W(u, u_1, \ldots, u_k)/W_{k+1} + (-1)^n W(u, u_1, \ldots, u_n)/W_n \qquad (26)$$

with continuous coefficients $c_k(t)$ $(k = 0, 1, \ldots, n-1)$.

LEMMA 17. *The coefficients* c_k *are given explicitly by the formulae*

$$c_0 = Lu_1$$

$$c_k = W_k^{-1} \sum_{j=1}^{k+1} (-1)^{j-1} W(u_1, \ldots, \hat{u}_j, \ldots, u_{k+1}) Lu_j \quad (k = 1, \ldots, n-1)$$

where the circumflex denotes that the corresponding term is to be omitted.

Proof. Taking $u = u_1, \ldots, u_{k+1}$ in (26) we obtain the system of linear equations

$$c_0 = Lu_1$$

$$c_0 u_2/u_1 - c_1 = Lu_2$$

$$c_0 u_3/u_1 + c_1 W(u_3, u_1)/W_2 + c_2 = Lu_3$$

$$\cdots\cdots$$

$$c_0 u_{k+1}/u_1 + c_1 W(u_{k+1}, u_1)/W_2 + \ldots + (-1)^k c_k = Lu_{k+1}.$$

Solving for c_k $(k \geq 1)$ by Cramer's rule we get

$$c_k = (-1)^{k(k+1)/2} \begin{vmatrix} 1 & 0 & 0 & \cdots & Lu_1 \\ u_2/u_1 & 0 & 0 & \cdots & Lu_2 \\ u_3/u_1 & W(u_3, u_1)/W_2 & 1 & \cdots & Lu_3 \\ & & \cdots\cdots & & \\ u_{k+1}/u_1 & W(u_{k+1}, u_1)/W_2 & W(u_{k+1}, u_1, u_2)/W_3 & \cdots & Lu_{k+1} \end{vmatrix}$$

$$= (W_1 W_2 \cdots W_k)^{-1} \begin{vmatrix} Lu_1 & u_1 & 0 & 0 & \cdots \\ Lu_2 & u_2 & W(u_1, u_2) & 0 & \cdots \\ Lu_3 & u_3 & W(u_1, u_3) & W(u_1, u_2, u_3) & \cdots \\ & & \cdots\cdots & & \\ Lu_{k+1} & u_{k+1} & W(u_1, u_{k+1}) & W(u_1, u_2, u_{k+1}) & \cdots \end{vmatrix} .$$

Adding u_1' times the second column to the third column we obtain

$$c_k = (W_2 \cdots W_k)^{-1} \begin{vmatrix} Lu_1 & u_1 & u_1' & 0 & \cdots \\ Lu_2 & u_2 & u_2' & 0 & \cdots \\ Lu_3 & u_3 & u_3' & W(u_1, u_2, u_3) & \cdots \\ & & \cdots\cdots & & \\ Lu_{k+1} & u_{k+1} & u_{k+1}' & W(u_1, u_2, u_{k+1}) & \cdots \end{vmatrix} .$$

Similarly, adding $-\begin{vmatrix} u_1' & u_2' \\ u_1'' & u_2'' \end{vmatrix}$ times the second column and $\begin{vmatrix} u_1 & u_2 \\ u_1'' & u_2'' \end{vmatrix}$ times the third column to the fourth column we obtain

$$c_k = (W_3 \cdots W_k)^{-1} \begin{vmatrix} Lu_1 & u_1 & u_1' & u_1'' & \cdots \\ Lu_2 & u_2 & u_2' & u_2'' & \cdots \\ Lu_3 & u_3 & u_3' & u_3'' & \cdots \\ & & \cdots\cdots & & \\ Lu_{k+1} & u_{k+1} & u_{k+1}' & u_{k+1}'' & \cdots \end{vmatrix} .$$

Proceeding in this manner we finally arrive at

$$c_k = W_k^{-1} \begin{vmatrix} Lu_1 & u_1 & u_1' & \cdots & u_1^{(k-1)} \\ Lu_2 & u_2 & u_2' & \cdots & u_2^{(k-1)} \\ Lu_3 & u_3 & u_3' & \cdots & u_3^{(k-1)} \\ & & \cdots\cdots & & \\ Lu_{k+1} & u_{k+1} & u_{k+1}' & \cdots & u_{k+1}^{(k-1)} \end{vmatrix} .$$

The expression for c_k assumes the required form if we now expand by the elements in the first column.

If we put

$$y_1 = u/u_1$$

$$y_k = W(u, u_1, \ldots, u_{k-1})/W_k \quad (k = 2, \ldots, n) ,$$

then

$$y_k' = - v_{k+1}y_{k+1} \quad (k = 1, \ldots, n-1) ,$$

$$y_n' = - W_{n-1}W(u, u_1, \ldots, u_n)/W_n^2 ,$$

where $v_{k+1} = W_{k-1}W_{k+1}/W_k^2 > 0$. Hence the equation (1) is equivalent to the system

$$\begin{aligned} y_k' &= - v_{k+1}y_{k+1} \quad (k = 1, \ldots, n-1) \\ y_n' &= (-1)^n \left[W_{n-1}/W_n\right] \sum_{k=0}^{n-1} c_k y_{k+1} . \end{aligned} \tag{27}$$

In this way any Markov system $u_1, \ldots, u_n$ can be used to replace the equation (1) by an equivalent system. We will be especially interested in Markov systems such that, in the corresponding system (27), $(-1)^n c_k \le 0$ $(k = 0, 1, \ldots, n-2)$. By Lemma 17 this will be the case if

$$W(u_1, \ldots, \hat{u}_j, \ldots, u_k) > 0 \quad (1 \le j < k < n)$$

and

$$(-1)^{n-k} Lu_k \ge 0 \quad (k = 1, \ldots, n-1) .$$

The first set of inequalities is certainly satisfied if $u_1, \ldots, u_{n-1}$ is a Descartes system.

If we compare the two expressions for y_n' obtained above we deduce the identity

$$(-1)^{n-1}W(u, u_1, \ldots, u_n)/W_n = c_0 u/u_1 + \sum_{k=1}^{n-1} c_k W(u, u_1, \ldots, u_k)/W_{k+1} .$$

This holds for any function u , any Markov system $u_1, \ldots, u_n$ and any n-th order linear differential operator L . In particular take $u = u_0$ to be any function such that $W(u_0, u_2, \ldots, u_n) \ne 0$ and let

$$Lu = (-1)^n W(u, u_0, u_2, \ldots, u_n)/W(u_0, u_2, \ldots, u_n) .$$

Then $Lu_j = 0$ $(j = 2, \ldots, n)$ and hence

$$c_0 = Lu_1 , \quad c_k = [W(u_2, \ldots, u_{k+1})/W_k]Lu_1 \quad (k = 1, \ldots, n-1) .$$

Since

$$Lu_1 = (-1)^{n-1} W(u_0, u_1, u_2, \ldots, u_n)/W(u_0, u_2, \ldots, u_n) ,$$

it follows that

$$W(u_0, u_2, \ldots, u_n)/W_n = u_0/u_1 + \sum_{k=1}^{n-1} W(u_2, \ldots, u_{k+1})W(u_0, u_1, \ldots, u_k)/W_k W_{k+1} . \quad (28)$$

This identity has been derived on the assumption that $W(u_0, u_2, \ldots, u_n) \neq 0$ but remains valid, by continuity, for arbitrary u_0 .

These formal relations will now be applied to questions of disconjugacy.

THEOREM 17. *Let* $u_1, \ldots, u_{n-1}$ *be a Descartes system on the interval* I *such that*

$$(-1)^{n-k} Lu_k \geq 0 \quad (k = 1, \ldots, n-1) .$$

Then the equation (1) *has a Markov fundamental system of solutions on* I .

Proof. Consider first the case of a compact interval $I = [a, b]$. We assume that $n > 1$ and the result holds for equations of order less than n . Choose u_n so that $W(u_1, \ldots, u_n) > 0$, *e.g.*, take u_n to be a solution of the equation $W(u_1, \ldots, u_{n-1}, u) = 1$. Then the equation (1) is equivalent to a system (27), where $(-1)^{n-1} c_k \geq 0$ $(k = 0, 1, \ldots, n-2)$. Thus if in (27) we change the independent variable from t to $-t$ we obtain a system of type K . If $y(t)$ is a solution of (27) such that $y(b) > 0$ it follows that $y(t) > 0$ for $a \leq t \leq b$. The corresponding solution $u = U_1$ of (1) satisfies

$$U_1 = u_1 y_1 > 0$$

$$W(U_1, u_1, \ldots, u_{k-1}) = W_k y_k > 0 \quad (k = 2, \ldots, n) .$$

Hence, by Proposition 4, $U_1, u_1, \ldots, u_{n-1}$ is a Descartes system on I . If we put $v = (u/U_1)'$ then $Lu \equiv Mv$, where M is a linear differential operator of order $n - 1$. Let

$$v_k = (u_{k+1}/U_1)' \quad (k = 1, \ldots, n-2) .$$

Since $U_1, u_2, \ldots, u_{n-1}$ is a Descartes system it follows from Lemma 3 that $v_1, \ldots, v_{n-2}$ is also a Descartes system. Moreover

$$(-1)^{n-1-k} Mv_k = (-1)^{n-k-1} Lu_{k+1} \geq 0 \quad (k = 1, \ldots, n-2) .$$

Therefore, by the induction hypothesis, the equation $Mv = 0$ has a Markov fundamental system of solutions $V_1, \ldots, V_{n-1}$. If we put

$$U_k(t) = U_1(t) \int_a^t V_{k-1}(s)ds \quad (k = 2, \ldots, n)$$

then, by Lemma 3 again, $U_1, \ldots, U_n$ is a Markov fundamental system of solutions of (1).

Suppose next that the interval I is open. Then, by what we have already proved, the equation (1) is disconjugate on any compact subinterval of I and hence on I itself. Therefore, by Theorem 15, it has a Markov fundamental system of solutions.

Suppose finally that $I = [a, b)$ is a half-open interval. Extend the definition of the coefficients of (1) by setting $p_k(t) = p_k(a)$ for $t < a$ and extend the definition of the functions u_k by setting $u_k(t) = \tilde{u}_k(t)$ for $t < a$, where $\tilde{u}_k$ is the solution of (1) which, together with its first $n - 1$ derivatives, takes the same value as u_k at $t = a$. Then the hypotheses of the theorem remain satisfied on an open interval (c, b), where c is less than and sufficiently close to a. Therefore the equation (1) has a Markov fundamental system of solutions on (c, b) and hence on $[a, b)$.

As an application we prove

PROPOSITION 16. *Suppose that for each* t *in the interval* I *the characteristic polynomial*

$$P_n(t, \lambda) = \lambda^n + p_1(t)\lambda^{n-1} + \ldots + p_n(t)$$

has real roots $\lambda_1(t), \ldots, \lambda_n(t)$ *which are separated by constants, i.e., there exist* $\mu_1 \leq \ldots \leq \mu_{n-1}$ *such that*

$$\lambda_1(t) \leq \mu_1 \leq \lambda_2(t) \leq \ldots \leq \mu_{n-1} \leq \lambda_n(t) . \tag{29}$$

Then the equation (1) *is disconjugate on* I.

Proof. The inequalities (29) imply that

$$(-1)^{n-k} P_n(t, \mu_k) \geq 0 \quad (k = 1, \ldots, n-1) .$$

If $\mu_1 < \mu_2 < \ldots < \mu_{n-1}$ then the functions $u_k(t) = e^{\mu_k t}$ $(k = 1, \ldots, n-1)$ are a Descartes system and

$$(-1)^{n-k} Lu_k = (-1)^{n-k} P_n(t, \mu_k) e^{\mu_k t} \geq 0 .$$

Therefore, by Theorem 17, the equation (1) is disconjugate on I. In particular this proves the proposition for $n = 2$.

Suppose $n > 2$ and the result holds for equations of order less than n. It only remains to consider the case where $\mu_k = \mu_{k+1}$ for some k. Then $\lambda_{k+1}(t) = \mu_k$ for all t and $e^{\mu_k t}$ is a solution of (1). If we put $v = \left\{e^{-\mu_k t} u\right\}'$ then $Lu \equiv Mv$, where M is a linear differential operator of order $n - 1$ with continuous coefficients and leading coefficient 1. Moreover the equation

$$Mv = 0 \tag{23}$$

has the characteristic polynomial $P_{n-1}(t, \lambda) = P_n(t, \lambda+\mu_k)/\lambda$. Evidently $P_{n-1}(t, \lambda)$ has the real roots

$$\lambda_1(t)-\mu_k, \ldots, \lambda_k(t)-\mu_k , \lambda_{k+2}(t)-\mu_k, \ldots, \lambda_n(t)-\mu_k ,$$

which are separated by the constants

$$\mu_1-\mu_k, \ldots, \mu_k-\mu_k = 0 , \mu_{k+2}-\mu_k, \ldots, \mu_{n-1}-\mu_k .$$

Therefore, by the induction hypothesis, the equation (23) is disconjugate on I. If a solution $u(t)$ of (1) has at least n zeros then, by Rolle's theorem, the solution $v(t) = \left[e^{-\mu_k t} u(t)\right]'$ of (23) has at least $n - 1$ zeros. Therefore $v(t) \equiv 0$ and $u(t) = \text{const.}e^{\mu_k t}$, which vanishes only if it is identically zero. This completes the proof.

To simplify the statement of our next result we will say that the functions $u_1, \ldots, u_n$ form a *weak Descartes system* on an interval I if all Wronskians $W\left(u_{i_1}, \ldots, u_{i_k}\right)$, where $1 \leq i_1 < \ldots < i_k \leq n$, are non-negative on I.

PROPOSITION 17. *Let I be an interval open at its right endpoint b and let $u_0, u_1, \ldots, u_n$ be C^n-functions on I satisfying the differential inequalities*

$$(-1)^{n-k} Lu_k \geq 0 \quad (k = 0, 1, \ldots, n) . \tag{30}$$

Suppose also that $u_1, \ldots, u_n$ is a Descartes system and $u_0, u_1, \ldots, u_n$ a weak

Descartes system.

Then the equation (1) *has a solution* U_1, *depending only on* u_1, *such that* $U_1, u_2, \ldots, u_n$ *is a Descartes system, while* $U_1, u_1, u_2, \ldots, u_n$ *and* $u_0, U_1, u_2, \ldots, u_n$ *are weak Descartes systems.*

Furthermore, the proposition remains valid if u_0 *and/or* u_n *is omitted in both hypothesis and conclusion.*

Proof. By Proposition 4 and its proof it is sufficient to show that

$$U_1 > 0 , \qquad W(U_1, u_2, \ldots, u_k) > 0 \quad (k = 2, \ldots, n) ,$$

$$W(U_1, u_1, \ldots, u_k) \geq 0 \quad (k = 1, \ldots, n) ,$$

$$W(u_0, U_1) \geq 0 , \quad W(u_0, U_1, u_2, \ldots, u_k) \geq 0 \quad (k = 2, \ldots, n) .$$

Let a_0 be a fixed point of I and let $J = [a, c]$ be a compact subinterval of I which contains a_0. We can choose $\lambda > 0$ so large that $e^{-\lambda t}, u_1, \ldots, u_n$ and $u_1, \ldots, u_{n-1}, e^{\lambda t}$ are Descartes systems on J and

$$(-1)^n L(e^{-\lambda t}) > 0 , \quad L(e^{\lambda t}) > 0 \quad \text{on } J .$$

If u_n is not given we take it to be $e^{\lambda t}$ in the argument which follows.

The equation (1) has a unique solution $U(t) = U(t, c)$ satisfying the initial conditions

$$u^{(k)} = u_1^{(k)} \quad (k = 0, 1, \ldots, n-1) \quad \text{at} \quad t = c .$$

By means of the functions $u_1, \ldots, u_n$ the equation (1) can be replaced by an equivalent system (27). The solution $y(t)$ of (27) which corresponds to the solution $U(t)$ of (1) has coordinates

$$y_1 = 1 , \quad y_k = 0 \quad (k = 2, \ldots, n) \quad \text{at} \quad t = c .$$

It follows, as in the proof of Theorem 12, that

$$y_1 > 0 , \quad y_k \geq 0 , \quad y_k' \leq 0 \quad (k = 2, \ldots, n) \quad \text{for} \quad a \leq t \leq c .$$

Thus

$$U > 0 , \quad W(U, u_1, \ldots, u_k) \geq 0 \quad (k = 1, \ldots, n) \quad \text{for} \quad a \leq t \leq c ,$$

and hence $U, u_1, \ldots, u_n$ is a weak Descartes system on J.

Put $\tilde{u}_1 = u_0 + \varepsilon e^{-\lambda t}$, where $\varepsilon > 0$. Then $\tilde{u}_1, u_2, \ldots, u_n$ form a Descartes system on J and in the same way we can use these functions to replace the equation (1) by an equivalent system (27). However, since $(-1)^{n-1}L\tilde{u}_1 \leq 0$, the previous argument requires modification. Using the identity (28) we see that the functions

$$y_2(t) = W(U, \tilde{u}_1)/W(\tilde{u}_1, u_2) ,$$

$$y_k(t) = W(U, \tilde{u}_1, u_2, \ldots, u_{k-1})/W(\tilde{u}_1, u_2, \ldots, u_k) \quad (k = 3, \ldots, n)$$

satisfy a system

$$y_k' = - v_{k+1}y_{k+1} \quad (k = 2, \ldots, n-1)$$

$$y_n' = \varphi - \sum_{k=2}^{n} b_k y_k ,$$

where

$$\varphi = (-1)^n L\tilde{u}_1 . W(U, u_0, \ldots, u_n) W(\tilde{u}_1, u_2, \ldots, u_{n-1}) / [W(\tilde{u}_1, u_2, \ldots, u_n)]^2$$
$$\geq 0$$

and the coefficients v_{k+1}, b_k are non-negative. Since $y_k(c) \leq 0 \quad (k = 2, \ldots, n)$ it follows from the theory of differential inequalities that $y_k(t) \leq 0$, $y_k'(t) \geq 0$ for $a \leq t \leq c \quad (k = 2, \ldots, n)$. Hence, throughout J,

$$W(\tilde{u}_1, U) \geq 0 , \quad W(\tilde{u}_1, U, u_2, \ldots, u_k) \geq 0 \quad (k = 2, \ldots, n) .$$

Letting $\varepsilon \to 0$ we obtain

$$W(u_0, U) \geq 0 , \quad W(u_0, U, u_2, \ldots, u_k) \geq 0 \quad (k = 2, \ldots, n) \quad \text{for} \quad a \leq t \leq c .$$

Now give a, c sequences of values a_ν, c_ν converging to the left and right endpoints of I. Also choose $\rho_\nu > 0$ so that

$$\rho_\nu^2 \sum_{k=0}^{n-1} \left[U^{(k)}(a_0, c_\nu)\right]^2 = 1 .$$

Then there exists a subsequence of $\{\rho_\nu U(t, c_\nu)\}$ which converges, together with its first n derivatives, uniformly on any compact subinterval of I. We can identify this subsequence with the original sequence. Then $U_1(t) = \lim_{\nu\to\infty} \rho_\nu U(t, c_\nu)$ is a nontrivial solution of (1). Moreover, for all t in I

$$U_1 \geq 0 , \qquad W(U_1, u_1, \ldots, u_k) \geq 0 \quad (k = 1, \ldots, n) ,$$

$$W(u_0, U_1) \geq 0 , \quad W(u_0, U_1, u_2, \ldots, u_k) \geq 0 \quad (k = 2, \ldots, n) .$$

The inequality $W(U_1, u_1) \geq 0$ implies that U_1/u_1 is non-increasing. Thus if $U_1(t_0) = 0$ then $U_1(t) = 0$ for $t \geq t_0$, which is a contradiction because U_1 is a nontrivial solution of (1). Therefore $U_1 > 0$ throughout I. We can now show by induction, as in the proof of Theorem 12, that $W(U_1, u_2, \ldots, u_k) > 0$ throughout I $(k = 2, \ldots, n)$.

Finally we show that the solution U_1 is independent of the sequence $\{c_\nu\}$ used to define it. In fact, let $\tilde{U}_1$ be the solution obtained from some other sequence. Since the conditions of the theorem remain satisfied if we replace u_0 by $\tilde{U}_1$ we have $W(\tilde{U}_1, U_1) \geq 0$. By symmetry, also $W(U_1, \tilde{U}_1) \geq 0$. Hence $\tilde{U}_1/U_1$ is a positive constant, which must be 1 because of the normalisation at $t = a_0$. Therefore

$$U_1(t) = \lim_{c \to b} U(t, c) .$$

THEOREM 18. *Let* $I = [a, b)$ *be a half-open interval and let* $u_0, u_1, \ldots, u_n$ *be* C^n*-functions on* I *satisfying the differential inequalities* (30). *Suppose also that* $u_1, \ldots, u_n$ *is a Descartes system and* $u_0, u_1, \ldots, u_n$ *a weak Descartes system.*

Then the equation (1) *has a fundamental system of solutions* $U_1, \ldots, U_n$ *which are positive and satisfy the inequalities*

$$u_0'/u_0 \leq U_1'/U_1 \leq u_1'/u_1 < U_2'/U_2 < \ldots < u_{n-1}'/u_{n-1} < U_n'/U_n < u_n'/u_n . \tag{31}$$

Moreover $U_1, \ldots, U_n$ *is a Markov system and*

$$W\left(U_1, \ldots, U_k, u_{i_k}, \ldots, u_{i_h}\right) \begin{cases} > 0 & \text{for } 1 \leq k < i_k < \ldots < i_h \leq n \\ \geq 0 & \text{for } 1 \leq k \leq i_k < \ldots < i_h \leq n , \end{cases} \tag{32}$$

$$\begin{gathered} W(U_1, \ldots, U_{k-1}, u_{k-1}, U_k) \geq 0 \quad \text{for } 1 \leq k \leq n \\ W\left(U_1, \ldots, U_{k-1}, u_{k-1}, U_k, u_{i_k}, \ldots, u_{i_h}\right) \geq 0 \quad \text{for } 1 \leq k < i_k < \ldots < i_h \leq n . \end{gathered} \tag{33}$$

The theorem remains valid if u_0 *and/or* u_n *is omitted in both hypothesis and conclusion.*

Proof. Since the result is trivial for $n = 1$ we assume that $n > 1$ and the result holds for equations of order less than n. Let U_1 be the solution whose existence was established in Proposition 17. If we put $v = (u/U_1)'$ then $Lu \equiv Mv$, where M is a linear differential operator of order $n-1$. Let

$$v_k = (u_{k+1}/U_1)' \quad (k = 0, 1, \ldots, n-1) .$$

It follows from Proposition 17 and Lemma 3 that $v_1, \ldots, v_{n-1}$ is a Descartes system and $v_0, v_1, \ldots, v_{n-1}$ a weak Descartes system. Moreover

$$(-1)^{n-1-k} Mv_k = (-1)^{n-k-1} Lu_{k+1} \geq 0 \quad (k = 0, 1, \ldots, n-1) .$$

Therefore, by the induction hypothesis, the equation $Mv = 0$ has a Markov fundamental system of positive solutions $V_1, \ldots, V_{n-1}$ such that

$$W(v_{k-1}, V_k) \geq 0 , \quad W(V_k, v_k) \geq 0 \quad (k = 1, \ldots, n-1) .$$

Moreover

$$W\left(V_1, \ldots, V_{k-1}, v_{i_k-1}, \ldots, v_{i_h-1}\right) \begin{cases} > 0 & \text{for } 1 < k < i_k < \ldots < i_h \leq n \\ \geq 0 & \text{for } 1 < k \leq i_k < \ldots < i_h \leq n , \end{cases}$$

$$W(V_1, \ldots, V_{k-2}, v_{k-2}, V_{k-1}) \geq 0 \quad \text{for } 1 < k \leq n ,$$

$$W\left(V_1, \ldots, V_{k-2}, v_{k-2}, V_{k-1}, v_{i_k-1}, \ldots, v_{i_h-1}\right) \geq 0 \quad \text{for } 1 < k < i_k < \ldots < i_h \leq n .$$

We can write $V_{k-1} = (U_k/U_1)'$, where U_k is a solution of (1). In fact

$$U_k = \tilde{U}_k + \alpha_k U_1 \quad (k = 2, \ldots, n) ,$$

where

$$\tilde{U}_k(t) = U_1(t) \int_a^t V_{k-1}(s)ds$$

and α_k is an arbitrary constant of integration. It follows from Lemma 3 that, for any choice of the constants α_k, the solutions $U_1, \ldots, U_n$ are a Markov system and the inequalities (32), (33) hold for $k > 1$. They hold also for $k = 1$ by Proposition 17. It remains to show that we can choose the constants $\alpha_k > 0$ $(k = 2, \ldots, n)$ so that the inequalities (31) involving $U_2, \ldots, U_n$ also hold.

For $1 < k \leq n$ we have

$$W(U_1, U_k, u_k) = U_1^3 W(V_{k-1}, v_{k-1}) \geq 0 ,$$

$$W(U_1, u_{k-1}, U_k) = U_1^3 W(v_{k-2}, V_{k-1}) \geq 0 .$$

From the identities

$$[W(U_k, u_k)/W(U_1, u_k)]' = u_k W(U_1, U_k, u_k)/W^2(U_1, u_k)$$

$$[W(u_{k-1}, U_k)/W(U_1, u_{k-1})]' = u_{k-1} W(U_1, u_{k-1}, U_k)/W^2(U_1, u_{k-1})$$

it follows that, if $k > 2$, the inequalities

$$W(U_k, u_k) > 0 , \quad W(u_{k-1}, U_k) > 0$$

hold for $t > a$ when they hold for $t = a$. But at $t = a$

$$\begin{aligned} W(U_k, u_k) &= W(\tilde{U}_k, u_k) + \alpha_k W(U_1, u_k) \\ &= - U_1 V_{k-1} u_k + \alpha_k W(U_1, u_k) \end{aligned}$$

and

$$\begin{aligned} W(u_{k-1}, U_k) &= W(u_{k-1}, \tilde{U}_k) - \alpha_k W(U_1, u_{k-1}) \\ &= U_1 V_{k-1} u_{k-1} - \alpha_k W(U_1, u_{k-1}) . \end{aligned}$$

Thus we wish to choose $\alpha_k > 0$ so that at $t = a$

$$- U_1 V_{k-1} u_{k-1} + \alpha_k W(U_1, u_{k-1}) < 0 < - U_1 V_{k-1} u_k + \alpha_k W(U_1, u_k) ,$$

that is,

$$U_1 V_{k-1} u_k / W(U_1, u_k) < \alpha_k < U_1 V_{k-1} u_{k-1} / W(U_1, u_{k-1}) .$$

This will be possible if and only if

$$W(U_1, u_{k-1})/u_{k-1} < W(U_1, u_k)/u_k ,$$

that is,

$$u'_{k-1}/u_{k-1} < u'_k/u_k .$$

However this is satisfied because $W(u_{k-1}, u_k) > 0$.

This argument remains valid for $k = 2$ unless $W(U_1, u_1)$ vanishes at some point of I. In this case let t_0 be the least value of t for which $W(U_1, u_1) = 0$. Then the proof of Proposition 17 shows that $W(U_1, u_1) = 0$ for $t_0 \le t < b$. Thus on this interval $u_1 = \beta U_1$ for some $\beta > 0$, and hence

$$W(u_1, U_2) = \beta W(U_1, U_2) > 0 .$$

On the other hand the previous argument enables us to choose $\alpha_2 > 0$ so that the inequality $W(U_2, u_2) > 0$ holds for $a \le t < b$ and the inequality $W(u_1, U_2) > 0$

holds for $a \le t < t_0$. This completes the proof.

By taking $u_k(t) = e^{\mu_k t}$ in Theorem 18 we obtain a stronger form of Proposition 16 under the stronger hypothesis $\mu_0 < \mu_1 < \ldots < \mu_n$.

PROPOSITION 18. *Suppose that for each* t *in the interval* $I = [a, \infty)$ *the characteristic polynomial*

$$P_n(t, \lambda) = \lambda^n + p_1(t)\lambda^{n-1} + \ldots + p_n(t)$$

has real roots $\lambda_1(t), \ldots, \lambda_n(t)$ *which are strictly separated by constants, i.e., there exist* $\mu_0 < \mu_1 < \ldots < \mu_n$ *such that*

$$\mu_0 \le \lambda_1(t) \le \mu_1 \le \ldots \le \mu_{n-1} \le \lambda_n(t) \le \mu_n .$$

Then the equation (1) *has a Markov fundamental system of positive solutions* $U_1, \ldots, U_n$ *which satisfy throughout* I

$$\mu_0 \le U_1'/U_1 \le \mu_1 < U_2'/U_2 < \ldots < \mu_{n-1} < U_n'/U_n < \mu_n ,$$

and hence

$$U_k(a)e^{\mu_{k-1}(t-a)} \le U_k(t) \le U_k(a)e^{\mu_k(t-a)} \quad (k = 1, \ldots, n) .$$

Again the proposition remains valid if μ_0 and/or μ_n is omitted in both hypothesis and conclusion. However if these hypotheses are retained then for the derivatives of the solutions $U_1, \ldots, U_n$ we have the estimates

$$U_k^{(i)} = O\left(e^{\mu_k t}\right) \quad (i, k = 1, \ldots, n) .$$

In fact, since the roots of the characteristic polynomial $P_n(t, \lambda)$ are bounded the coefficients are also bounded. If in (1) we make the change of variables $u = e^{\mu_k t} v$ then the transformed equation with leading coefficient 1 again has bounded coefficients. The solution U_k of (1) is transformed into a bounded solution V_k of the new equation. But this implies (see, *e.g.*, Coppel[1], p. 141), that the derivatives $V_k^{(i)}$ $(i = 1, \ldots, n)$ are also bounded. Hence

$$U_k^{(i)} = \left(e^{\mu_k t} V_k\right)^{(i)} = O\left(e^{\mu_k t}\right) .$$

We prove next a simple lemma on differential inequalities.

LEMMA 18. *Suppose the equation* (1) *is disconjugate on the interval* I *with right endpoint* b . *If* $L\tilde{u} \geq 0$ [*or* $L\tilde{u} \leq 0$] *on* I *then either* $\tilde{u}(t) \equiv 0$ *or* $\tilde{u}(t) \neq 0$ *in the neighbourhood of* b .

Moreover if U *is any nontrivial solution of* (1) *then* $U(t) \neq 0$ *and the ratio* $\tilde{u}(t)/U(t)$ *is monotonic in the neighbourhood of* b .

Proof. Suppose first that $n = 1$. From the explicit formula for the solution of an inhomogeneous first order linear equation or, alternatively, by the theory of differential inequalities if $\tilde{u}(t_1) = \tilde{u}(t_2) = 0$ then $\tilde{u}(t) = 0$ for $t_1 \leq t \leq t_2$. Hence if $\tilde{u}(t_\nu) = 0$ for a sequence $t_\nu \to b$ then $\tilde{u}(t) \equiv 0$ near b .

Suppose next that the first statement holds for equations of order less than n . If U is a nontrivial solution of (1) it has no zeros near b . Putting $v = (u/U)'$ we obtain in the usual way $Lu \equiv Mv$. Moreover the equation $Mv = 0$ is disconjugate on I . If $\tilde{v} = (\tilde{u}/U)'$ then $M\tilde{v} = L\tilde{u} \geq 0$. Hence, by the induction hypothesis, either $\tilde{v}(t) \equiv 0$ or $\tilde{v}(t) \neq 0$ near b . Thus $\tilde{u}/U$ is either constant or strictly monotonic near b . Hence either $\tilde{u}(t) \equiv 0$ or $\tilde{u}(t) \neq 0$ near b .

We are going to use this lemma to study the asymptotic behaviour of the solutions of an equation near an endpoint of the interval of disconjugacy. We will require in addition the following extension of Proposition 14 in which the Wronskian $W(u_1, \ldots, u_n)$ is allowed to vanish at some points.

PROPOSITION 19. *Let* I *be an interval open at its right endpoint* b *and let* $u_1, \ldots, u_n$ *be* C^n*-functions which are positive near* b *and satisfy* $u_{k-1} = o(u_k)$ *for* $t \to b$ $(k = 2, \ldots, n)$.

If $W(u_1, \ldots, u_k) \neq 0$ $(k = 1, \ldots, n-1)$ *and* $W(u_1, \ldots, u_n)$ *does not change sign on* I *then* $u_1, \ldots, u_{n-1}$ *is a Markov system and* $W(u_1, \ldots, u_n) \geq 0$ *on* I , *while near* b

$$W\left(u_{i_1}, \ldots, u_{i_k}\right) > 0 \quad \textit{for} \quad 1 \leq i_1 < \ldots < i_k \leq n \quad (1 \leq k < n) . \qquad (34)$$

If $W(u_k, \ldots, u_n) \neq 0$ $(k = 2, \ldots, n)$ *and* $W(u_1, \ldots, u_n)$ *does not change sign on* I *then* $W(u_1, \ldots, u_n) \geq 0$ *and* (34) *hold throughout* I .

Proof. Suppose first that $W(u_1, \ldots, u_k) \neq 0$ on I $(k = 1, \ldots, n-1)$. Then $u_1, \ldots, u_{n-1}$ form a Markov system on I and a Descartes system near b , by Proposition 14. Since all solutions u of the equation $W(u_1, \ldots, u_{n-1}, u) = 0$ satisfy $u = o(u_n)$ we cannot have $W(u_1, \ldots, u_{n-1}, u_n) \equiv 0$ near b . It follows from (24), as in the proof of Proposition 14, that $W(u_k, \ldots, u_n) \neq 0$ near b

$(k = 2, \ldots, n)$. Hence, by Proposition 14 again, $u_2, \ldots, u_n$ is a Descartes system near b . It now follows from the proof of Proposition 4 that (34) holds near b .

If $W(u_1, \ldots, u_n)$ is non-positive on I then in the neighbourhood of b

$$[W(u_2, \ldots, u_n)/W(u_1, \ldots, u_{n-1})]' \leq 0 .$$

Hence there exists γ $(0 < \gamma < \infty)$ such that

$$W(u_2, \ldots, u_n) < \gamma W(u_1, \ldots, u_{n-1}) ,$$

that is

$$\tilde{W} = W(u_2, \ldots, u_{n-1}, \tilde{u}_n) < 0 ,$$

where $\tilde{u}_n = u_n + (-1)^{n-1}\gamma u_1$. Thus $u_2, \ldots, u_{n-1}, \tilde{u}_n$ are solutions satisfying [B] of an equation of order $n - 1$ which has the Markov fundamental system of solutions $u_2, \ldots, u_{n-1}, -\tilde{u}_n$. Therefore, by Theorem 14, $\tilde{W} > 0$, which is a contradiction. We conclude that $W(u_1, \ldots, u_n) > 0$ on I .

Suppose next that $W(u_k, \ldots, u_n) \neq 0$ on I $(k = 2, \ldots, n)$. Then $u_2, \ldots, u_n$ is a Descartes system on I , by Proposition 14. Since all nontrivial solutions u of the equation $W(u_2, \ldots, u_n, u) = 0$ satisfy $u_1 = o(u)$ we cannot have $W(u_1, u_2, \ldots, u_n) \equiv 0$ near b . It follows from (25) that $W(u_1, \ldots, u_k) \neq 0$ near b $(k = 1, \ldots, n-1)$. Hence, by what we have already proved, $W(u_1, \ldots, u_n) \geq 0$ throughout I and (34) holds near b . If (34) does not hold throughout I there exists $a \in I$ such that (34) holds on (a, b) but not on $[a, b)$. Thus there exists a k-tuple $i_1 < \ldots < i_k$ $(1 \leq k < n)$ with $i_1 = 1$ such that $W\left(u_{i_1}, \ldots, u_{i_k}\right) = 0$ at $t = a$. We can choose a k-tuple $j_1 < \ldots < j_k$ with $j_1 = 2$ such that $i_\alpha \leq j_\alpha$ $(\alpha = 1, \ldots, k)$. Then by the proof of Theorem 14 the ratio

$$W\left(u_{i_1}, \ldots, u_{i_k}\right) \Big/ W\left(u_{j_1}, \ldots, u_{k}\right)$$

is a non-increasing function on (a, b) . Since it is positive near b this gives a contradiction.

THEOREM 19. *Let I be an interval with right endpoint b and let $u_0, u_1, \ldots, u_n$ be C^n-functions on I which are positive near b and satisfy*

$$(-1)^{n-k}Lu_k \geq 0 \quad on \quad I \quad (k = 0, \ldots, n) \, ,$$

$$u_{k-1} = o(u_k) \quad for \quad t \to b \quad (k = 1, \ldots, n) \, ,$$

$$W(u_1, \ldots, u_k) \neq 0 \quad near \quad b \quad (k = 1, \ldots, n-1) \, .$$

Then the equation (1) *has a Markov fundamental system of solutions near* b *which satisfies* [B]. *Moreover if* $U_1, \ldots, U_n$ *are any solutions satisfying* [B] *then for* $t \to b$

$$U_k = O(u_k) \, , \qquad (35)$$

$$(k = 1, \ldots, n) \, .$$

$$u_{k-1} = O(U_k) \, . \qquad (36)$$

Proof. The functions $u_1, \ldots, u_{n-1}$ are a Descartes system on a subinterval $J = [a, b)$, by Proposition 14. Hence, by Theorems 17 and 13, the equation (1) has a Markov fundamental system of solutions $U_1, \ldots, U_n$ on J which satisfies [B]. The inequality $(-1)^{n-k}Lu_k \geq 0$ can be written

$$(-1)^{n-k}W(U_1, \ldots, U_n, u_k) \geq 0 \, . \qquad (37)_k$$

In particular $W(U_1, \ldots, U_n, u_k)$ is of constant sign.

We show first that $u_{n-1} = O(U_n)$. If this is not the case then, by Lemma 18, $U_n = o(u_{n-1})$. Thus the functions $U_1, \ldots, U_n, u_{n-1}$ satisfy the conditions of Proposition 19 and consequently $W(U_1, \ldots, U_n, u_{n-1}) \geq 0$ near b . This contradicts $(37)_{n-1}$ unless u_{n-1} is a solution of the equation (1) near b . But this is impossible, because $u = o(u_n)$ for all solutions u of (1).

We show next that $u_k = O(U_{k+1})$ implies $u_{k-1} = O(U_k)$. If this is not the case then $U_k = o(u_{k-1})$. Since $u_{k-1} = o(u_k) = o(U_{k+1})$ it follows that u_{k-1} is not a solution of (1) near b . Hence - *cf.* the proof of Proposition 19 - $W(U_1, \ldots, U_{n-1}, u_{k-1}) \neq 0$ near b . Thus the equation $W(U_1, \ldots, U_{n-1}, u_{k-1}, u) = 0$ has the Markov fundamental system of solutions $U_1, \ldots, U_{n-1}, \pm u_{k-1}$. Since the solutions $U_1, \ldots, U_k, u_{k-1}, U_{k+1}, \ldots, U_{n-1}$ satisfy [B] they form a Markov system. Applying Proposition 19 to the functions $U_1, \ldots, U_k, u_{k-1}, U_{k+1}, \ldots, U_n$ we obtain $W(U_1, \ldots, U_k, u_{k-1}, U_{k+1}, \ldots, U_n) \geq 0$ near b , which contradicts $(37)_{k-1}$. This proves (36).

The proof of (35) is similar. We show first that $U_1 = O(u_1)$. If this is not the case then $u_1 = o(U_1)$. Thus u_1 is not a solution of (1) near b . Applying Proposition 19 to the functions $u_1, U_1, \ldots, U_n$ we obtain $W(u_1, U_1, \ldots, U_n) \geq 0$, which contradicts $(37)_1$. Moreover, in the same way as before we can show that $U_{k-1} = O(u_{k-1})$ implies $U_k = O(u_k)$.

The proof of the existence of a Markov fundamental system on J was based on Theorem 17 and thus did not involve u_0 or u_n . Also the estimates $u_0 = O(U_1)$, $U_n = O(u_n)$ were the final links in their inductive chains and thus were not used in obtaining the other estimates. It follows that also Theorem 19 remains valid if u_0 and/or u_n is omitted in both hypothesis and conclusion.

NOTES

Chapter 0. The term 'disconjugate' is due to Wintner [2]. In Russian and in older American literature 'nonoscillatory' is often used instead. This differs from our use of 'nonoscillatory'.

Chapter 1. Oscillation theory has its origin in the basic work of Sturm [1]. Related eigenfunction expansions were treated by Liouville [1]. A connected account of the theory for second order equations is given by Bôcher [3].

Lemma 1 appears in Leighton and Nehari [1]. The significance of Riccati equations and Riccati inequalities for questions of disconjugacy was pointed out by Bôcher [1], [2]. For Theorems 3 and 4 see also de la Vallée Poussin [1]. Principal solutions were studied by Morse and Leighton [1] and by Hartman and Wintner [1]; the name is due to Leighton [1].

For a discussion of the Jacobi condition in the calculus of variations see *e.g.*, Gelfand and Fomin [1]. The identity of Lemma 2 is the basis of the so-called Legendre transformation of the second variation. For Proposition 3 *cf.* Leighton [2]. The identity (12) was given by Picone [1]. It forms the basis for the treatment of comparison theory in Ince [1]. The method of polar coordinates was used for the same purpose by Kamke [1]. Proposition 5 is due to Barrett [1]. Hartman has pointed out that by a change of independent variable the proposition can be deduced from the special case $p(t) \equiv 1$, $b = \infty$, for which a simple convexity argument suffices. The first integral comparison test was proved by Hille [1]. The more general treatment here follows Taam [1].

For Theorem 11 and Proposition 6 see Rab [1] and Wintner [1] respectively. Theorem 12 is contained in Moore [1]. The sequence of tests obtained by repeated use of the transformation (19) is given by Miller [1]. Theorem 13 is due to Liapounoff [1], who used it to discuss the stability of Hill's equation. The proof given here is due to Nehari [1]; the proof that the constant 4 is best possible follows Borg [2]. An analogue of Liapounoff's result for the L^2-norm has been obtained by Borg [1].

The concept of convergence in L^2 , which is introduced here, serves a similar purpose to the approximate convergence of Olech, Opial and Wazewski [1]. Propositions 8 and 9 are due to Hartman [2]. For Theorem 14 see Markus and Moore [1]. The replacement of the Riccati equation by an integral equation occurs in Hille [1]. Theorem 15 is proved by Opial [1]. Theorem 16 is a reformulation of a result of Hartman [1]. For Theorems 17 and 18 see Willett [1] and Wong [1]. Further results are given in Hartman [5], Swanson [1] and Willett [2].

Chapter 2. Our isotropic solutions are usually called conjugate or conjoined solutions. This seems to be overworking the word 'conjugate'. The method of proof of Theorem 1 is due to Hestenes; see Reid [2]. The concept of controllability developed from the classical normality condition of the calculus of variations. For its role in control theory see Kalman *et al.* [1].

Principal solutions for systems of equations have been considered by Hartman [3], [5] and by Reid [4], [5]. Conjugate points were treated by Radon [1], where matrix Riccati equations also appear. The latter are further discussed by Reid [1], [5] and Sandor [1]. Some of the results for matrix Riccati equations given here are new. We have restricted attention to the self-adjoint type which arises from linear Hamiltonian systems. However some results can also be established for the general equation

$$W' + A + WB + DW + WCW = 0 ,$$

without any symmetry assumptions on the coefficients. For example (Reid [5]), if W_1, W_2, W_3, W_4 are four solutions such that $W_3 - W_2$ and $W_4 - W_1$ are invertible on an interval I then their cross-ratio

$$\{W_1, W_2, W_3, W_4\} = (W_3 - W_1)(W_3 - W_2)^{-1}(W_4 - W_2)(W_4 - W_1)^{-1}$$

is similar to a constant matrix. This generalises a well-known property of scalar Riccati equations.

The formulation of the variational problem follows Reid [5]. This formulation is suggested by the theory of optimal control; see *e.g.*, Pontryagin *et al.* [1], Lee and Markus [1]. The extension to systems of the Legendre transformation of the second variation is due to Clebsch [1]. The proof of Proposition 11 follows Radon [1]. The first extensions of the Sturm comparison theorem to systems are due to Morse [1] and Birkhoff and Hestenes [1]. For Theorem 15 and Proposition 13 see Reid [3], [2]. For the extension of the polar coordinate method to systems see Lidskiĭ [1] and Atkinson [1]. Oscillation properties of linear Hamiltonian systems have been studied by Jakubovič [1]. A basic result of Morse connects the number of conjugate points with the number of negative eigenvalues of a corresponding variational problem; see Morse [2], Milnor [1], Hestenes [1].

The concept of adjoint equation goes back to Lagrange. For Theorems 17 - 19 see Krein [1] and Heinz [1]. Connections with spectral theory are studied by Glazman [1] and Ladas [1].

Chapter 3. Proposition 3 is a particular case of a result for nonlinear equations due to Hartman [4]. The present proof follows Opial [2].

A simple disconjugacy criterion for n-th order equations was proved by Fite [1]. His result was improved by de la Vallée Poussin [1]. Theorem 1, which is due to Levin [2], is a further improvement. Hartman [7] has an even sharper, but more complicated, result which he uses to obtain a valid proof of a disconjugacy criterion of Nehari [2]. Bessmertnyh and Levin [1] obtained the criterion

$$\sum_{k=1}^{n-1} \frac{n-k}{k!n} M_k(b-a)^k + \frac{(n-1)^{n-1}}{n!n^n} M_n(b-a)^n < 1 ,$$

which can give either stronger or weaker results than Levin's criterion. See also Levin [3].

Čebyšev systems play a basic role in approximation theory; see, *e.g.*, Cheney [1]. Our definition is slightly different from the usual one because we count zeros according to their multiplicity. However Propositions 3 and 12 show that for sufficiently smooth functions and open or half-open intervals the two definitions coincide. The definitions of Markov and Descartes systems are also modifications of those usual in approximation theory. Proposition 4 is contained in a purely algebraic result due to Fekete; see Gantmakher and Krein [1]. For Proposition 6 see Pólya and Szegö [1]. The factorisation of linear differential operators has been studied by Frobenius [1], Pólya [1] and Mammana [1]. Mammana conjectured that any operator of odd order has a linear factor. However this is disproved by examples of third order equations for which every solution has at least one zero on a given compact interval; see Sansone [1].

The basic properties of conjugate points and adjoint equations were stated by Levin [1]. The present treatment follows Hinton [1]. The sign of the Green's function was determined by Levin [1] and Pokornyi [1]. For Proposition 13 *cf.* Čičkin [1] and Hartman [4]. It has been shown by Krein [2] that if the boundary conditions are imposed at the endpoints only then the Green's function is *totally positive* and that, conversely, this property characterises disconjugate equations. For the definition of total positivity see Karlin [1]. Total positivity of the Green's function has important consequences for the corresponding eigenfunction expansions; see Krein [3] and Gantmakher and Krein [1].

The extension of the notion of principal solution to equations of arbitrary order is quite recent. The discussion in Sections 7 and 8 is based on the noteworthy papers of Hartman [6] and Levin [3]. Levin gives many interesting applications of Theorem 17. Theorem 18, which was communicated by Hartman, is an improved and corrected form of Theorem 11.1 in his paper. The relation between the oscillation properties of a third-order equation and those of its adjoint is studied by Dolan [1]. For the analogue of Theorem 11 of Chapter 2 see Sherman [1].

REFERENCES

F.V. Atkinson:

[1] *Discrete and continuous boundary problems* (Academic Press, New York, London, 1964). [MR31#416].

John H. Barrett:

[1] "Disconjugacy of second-order linear differential equations with nonnegative coefficients", *Proc. Amer. Math. Soc.* 10 (1959), 552-561. [MR21#7329].

[2] "Oscillation theory of ordinary linear differential equations", *Advances in Math.* 3 (1969), 415-509. [MR41#2113].

G.A. Bessmertnyh and A.Ju. Levin:

[1] "О некоторых оценках дифференцируемых функций одной переменной", *Dokl. Akad. Nauk SSSR* 144 (1962), 471-474. [MR25#1245]. "Some inequalities satisfied by differentiable functions of one variable", *Soviet Math. Dokl.* 3 (1962), 737-740.

G.D. Birkhoff and M.R. Hestenes:

[1] "Natural isoperimetric conditions in the calculus of variations", *Duke Math. J.* 1 (1935), 198-286. [Zbl. Math. 14, 316]. [See also *Proc. Nat. Acad. Sci. USA* 21 (1935), 99-102.]

Maxime Bôcher:

[1] "Application of a method of d'Alembert to the proof of Sturm's theorems of comparison", *Trans. Amer. Math. Soc.* 1 (1900), 414-420.

[2] "Non-oscillatory linear differential equations of the second order", *Bull. Amer. Math. Soc.* 7 (1900-01), 333-340.

[3] *Leçons sur les méthodes de Sturm* (Gauthier-Villars, Paris, 1917).

Göran Borg:

[1] "Über die Stabilität gewisser Klassen von linearen Differentialgleichungen", *Ark. Mat. Astr. Fys.* 31A, no. 1 (1944). [MR8, 70].

[2] "On a Liapounoff criterion of stability", *Amer. J. Math.* 71 (1949), 67-70. [MR10, 456].

E.W. Cheney:

[1] *Introduction to approximation theory* (McGraw-Hill, New York, St Louis, San Francisco, Toronto, London, Sydney, 1966). [MR36#5568].

E.S. Čičkin:

[1] "Теорема о дифференциальном неравенстве для многоточечных краевых задач", *Izv. Vysš. Učebn. Zaved. Matematika* 2 (27) (1962), 170-179. [MR26#1541]. [= "A theorem on a differential inequality for multipoint boundary value problems"]

A. Clebsch:

[1] "Ueber die Reduktion der Zweiten Variation auf ihre einfachste Form", *J. reine angew. Math.* 55 (1858), 254-273.

W.A. Coppel:

[1] *Stability and asymptotic behavior of differential equations* (D.C. Heath, Boston, 1965). [MR32#7875].

J.M. Dolan:

[1] "On the relationship between the oscillatory behavior of a linear third-order differential equation and its adjoint", *J. Differential Equations* 7 (1970), 367-388. [MR41#568].

J. Favard:

[1] "Sur les équations différentielles linéaires à coefficients presque-périodiques", *Acta Math.* 51 (1928), 31-81.

William Benjamin Fite:

[1] "Concerning the zeros of the solutions of certain differential equations", *Trans. Amer. Math. Soc.* 19 (1918), 341-352.

G. Frobenius:

[1] "Ueber die Determinante mehrerer Funktionen einer Variabeln", *J. reine angew. Math.* 77 (1874), 245-257. [= *Gesammelte Abhandlungen*, Band 1 (1968), 141-153.]

F.R. Gantmacher; M.G. Krein:

[1] *Oszillationsmatrizen, Oszillationskerne und kleine Schwingungen mechanischer Systeme* (Akademie-Verlag, Berlin, 1960). [MR22#5161].

I.M. Gelfand and S.V. Fomin:

[1] *Calculus of variations* (Russian; translated by R.A. Silverman; Prentice-Hall, Englewood Cliffs, New Jersey, 1963). [MR28#3353].

I.M. Glazman:

[1] *Direct methods of qualitative spectral analysis of singular differential operators* (Russian; Gosudarstv. Izdat. Fiz.-Mat. Lit., Moscow, 1963). [MR32#2938]. (English transl., Israel Program for Scientific Translations, Jerusalem, 1965.)

Philip Hartman:

[1] "On linear second order differential equations with small coefficients", *Amer. J. Math.* 73 (1951), 955-962. [MR13, 652].

[2] "On non-oscillatory linear differential equations of second order", *Amer. J. Math.* 74 (1952), 389-400. [MR14, 50].

[3] "Self-adjoint, non-oscillatory systems of ordinary, second order, linear differential equations", *Duke Math. J.* 24 (1957), 25-35. [MR18, 576].

[4] "Unrestricted n-parameter families", *Rend. Circ. Mat. Palermo* (2) 7 (1958), 123-142. [MR21#4211].

[5] *Ordinary differential equations* (John Wiley & Sons, New York, London, Sydney, 1964). [MR30#1270].

[6] "Principal solutions of disconjugate n-th order linear differential equations", *Amer. J. Math.* 91 (1969), 306-362. [MR40#450].

[7] "On disconjugacy criteria", *Proc. Amer. Math. Soc.* 24 (1970), 374-381. [Zbl. Math. 191, 383].

Philip Hartman and Aurel Wintner:

[1] "On the assignment of asymptotic values for the solutions of linear differential equations of second order", *Amer. J. Math.* 77 (1955), 475-483. [MR17, 36].

Erhard Heinz:

[1] "Halbbeschränktheit gewöhnlicher Differentialoperatoren höherer Ordnung", *Math. Ann.* 135 (1958), 1-49. [MR21#743].

Magnus R. Hestenes:

[1] "Applications of the theory of quadratic forms in Hilbert space to the calculus of variations", *Pacific J. Math.* 1 (1951), 525-581. [MR13, 759].

Einar Hille:

[1] "Non-oscillation theorems", *Trans. Amer. Math. Soc.* 64 (1948), 234-252. [MR10, 376].

Don B. Hinton:

[1] "Disconjugate properties of a system of differential equations", *J. Differential Equations* 2 (1966), 420-437. [MR34#7856].

E.L. Ince:

[1] *Ordinary differential equations* (Longmans, London, 1927).

V.A. Jakubovič:

[1] "Осцилляторные свойства решений канонических уравнений", *Mat. Sb. (N.S.)* 56 (98) (1962), 3-42. [MR25#2303]. "Oscillatory properties of solutions of canonical equations", *Amer. Math. Soc. Transl.* (2) 42 (1964), 247-288.

R.E. Kalman, Y.C. Ho and K.S. Narendra:

[1] "Controllability of linear dynamical systems", *Contributions to Differential Equations* 1 (1963), 189-213. [MR27#5012].

E. Kamke:

[1] "A new proof of Sturm's comparison theorems", *Amer. Math. Monthly* 46 (1939), 417-421. [MR1, 54].

Samuel Karlin:

[1] *Total positivity*, vol. 1 (Stanford University Press, Stanford, California, 1968). [MR37#5667].

M.G. Krein:

[1] "Sur les opérateurs différentiels autoadjoints et leurs fonctions de Green symétriques", *Mat. Sb. (N.S.)* 2 (44) (1937), 1023-1072. [Zbl. Math. 18, 257].

[2] "Sur les fonctions de Green non-symétriques oscillatoires des opérateurs différentiels ordinaires", *C.R. (Doklady) Acad. Sci. URSS (N.S.)* 25 (1939), 643-646. [MR2, 52].

[3] "Les théorèmes d'oscillation pour les opérateurs linéaires différentiels d'ordre quelconque", *C.R. (Doklady) Acad. Sci. URSS (N.S.)* 25 (1939), 719-722. [MR2, 53].

Gerasimos Ladas:

[1] "Connection between oscillation and spectrum for selfadjoint differential operators of order $2n$ ", *Comm. Pure Appl. Math.* 22 (1969), 561-585. [Zbl. Math. 93, 46].

E.B. Lee; L. Markus:

[1] *Foundations of optimal control theory* (John Wiley & Sons, New York, London, Sydney, 1967). [MR36#3596].

Walter Leighton:

[1] "Principal quadratic functionals", *Trans. Amer. Math. Soc.* 67 (1949), 253-274. [MR11, 603].

[2] "Comparison theorems for linear differential equations of second order", *Proc. Amer. Math. Soc.* 13 (1962), 603-610. [MR25#4173].

Walter Leighton and Zeev Nehari:

[1] "On the oscillation of solutions of self-adjoint linear differential equations of the fourth order", *Trans. Amer. Math. Soc.* 89 (1958), 325-377. [MR21#1429].

A. Ju. Levin:

[1] "Некоторые вопросы осцилляции решений линейных дифференциальных уравнений", *Dokl. Akad. Nauk SSSR* 148 (1963), 512-515. [MR26#3972]. "Some problems bearing on the oscillation of solutions of linear differential equations", *Soviet Math. Dokl.* 4 (1963), 121-124.

[2] "Оценка для функции с монотонно расположенными нулями последовательных производных", *Mat. Sb.* 64 (106) (1964), 396-409. [MR29#4941]. [= "A bound for a function with monotonely distributed zeros of successive derivatives"].

[3] "Неосцилляция решений уравнения $x^{(n)} + p_1(t)x^{(n-1)} + \ldots + p_n(t)x = 0$", *Uspehi Mat. Nauk* 24 (1969), 43-96. [MR40#7537]. "Non-oscillation of solutions of the equation $x^{(n)} + p_1(t)x^{(n-1)} + \ldots + p_n(t)x = 0$", *Russian Math. Surveys* 24 (1969), 43-99.

M.A. Liapounoff:

[1] *Problème général de la stabilité du mouvement* (Annals of Math. Studies, no. 17, Princeton University Press, Princeton; Oxford University Press, London; 1947). [MR9, 34].

V.B. Lidskiĭ:

[1] "Осцилляционные теоремы для канонической системы дифференциальных уравнений", *Dokl. Akad. Nauk SSSR* 102 (1955), 877-880. [MR17, 483]. [= "Oscillation theorems for canonical systems of differential equations"].

J. Liouville:

[1] "Sur le développement des fonctions en séries dont divers termes sont assujettis à satisfaire à une même équation différentielle du second ordre contenant un paramètre variable", *J. Math. Pures Appl.* 1 (1836), 253-265 and 2 (1837), 16-35.

Gabriele Mammana:

[1] "Decomposizione delle espressioni differenziale lineari omogenee in prodotti di fattori simbolici e applicazione relativa allo studio delle equazioni differenziali lineari", *Math. Z.* 33 (1931), 186-231. [Zbl. Math. 1, 15].

Laurence Markus and Richard A. Moore:

[1] "Oscillation and disconjugacy for linear differential equations with almost periodic coefficients", *Acta Math.* 96 (1956), 99-123. [MR18, 306].

J.C.P. Miller:

[1] "On a criterion for oscillatory solutions of a linear differential equation of the second order", *Proc. Cambridge Philos. Soc.* 36 (1940), 283-287. [MR2, 50].

J. Milnor:

[1] *Morse theory* (Based on lecture notes by M. Spivak and R. Wells. Annals of Math. Studies, no. 51, Princeton University Press, Princeton, New Jersey, 1963). [MR29#634].

Richard A. Moore:

[1] "The behavior of solutions of a linear differential equation of second order", *Pacific J. Math.* 5 (1955), 125-145. [MR16, 925].

Marston Morse:

[1] "A generalization of the Sturm separation and comparison theorems in n-space", *Math. Ann.* 103 (1930), 52-69.

[2] *The calculus of variations in the large* (Amer. Math. Soc. Colloq. Publ. vol. 18, Amer. Math. Soc., New York, 1934). [Zbl. Math. 11, 28].

Marston Morse and Walter Leighton:

[1] "Singular quadratic functionals", *Trans. Amer. Math. Soc.* 40 (1936), 252-286. [Zbl. Math. 15, 27].

Zeev Nehari:

[1] "On the zeros of solutions of second-order linear differential equations", *Amer. J. Math.* 76 (1954), 689-697. [MR16, 131].

[2] "On an inequality of Lyapunov", *Studies in mathematical analysis and related topics*, 256-261. (Stanford University Press, Stanford, California, 1962.) [MR26#2684].

C. Olech, Z. Opial et T. Ważewski:

[1] "Sur le problème d'oscillation des intégrales de l'équation $y'' + g(t)y = 0$ ", *Bull. Acad. Polon. Sci. Cl. III* 5 (1957), 621-626. [MR19, 650].

Zdzisław Opial:

[1] "Sur les intégrales oscillantes de l'équation différentielle $u'' + f(t)u = 0$ ", *Ann. Polon. Math.* 4 (1958), 308-313. [MR20#4051].

[2] "On a theorem of O. Aramă", *J. Differential Equations* 3 (1967), 88-91. [MR34#6194].

M. Picone:

[1] "Sui valori eccezionali di un parametro da cui dipende un'equazione differenziale lineare del secondo ordine", *Ann. Scuola Norm. Sup. Pisa* (1) 11 (1910), 1-141. [Revue semestrielle 23, 2^e partie, 61.]

Ju.V. Pokornyĭ:

[1] "О некоторых оценках функции Грина многоточечной краевой задачи", *Mat. Zametki* 4 (1968), 533-540. [MR38#4748]. [= "Some estimates of the Green's function of a multi-point boundary value problem".]

G. Pólya:

[1] "On the mean-value theorem corresponding to a given linear homogeneous differential equation", *Trans. Amer. Math. Soc.* 24 (1924), 312-324.

G. Pólya und G. Szegö:

[1] *Aufgaben und Lehrsätze aus der Analysis*, Erster Band, 2te Aufl. (Die Grundlehren der mathematischen Wissenschaften in Einzeldarstellungen, Band 19, Springer-Verlag, Berlin, Göttingen, Heidelberg, 1954). [MR15, 512].

L.S. Pontrjagin, V.G. Boltjanskiĭ, R.V. Gamkrelidze and E.F. Miščenko:

[1] *The mathematical theory of optimal processes* (translated by K.N. Trirogoff; edited by L.W. Neustadt. Interscience [John Wiley & Sons], New York, London, 1962). [MR29#3316b].

Miloš Ráb:

[1] "Kriterien für die Oszillation der Lösungen der Differentialgleichung $[p(x)y']' + q(x)y = 0$ ", *Časopis Pěst. Mat.* 84 (1959), 335-370 and 85 (1960), 91. [MR22#5773].

Johann Radon:

[1] "Zum Problem von Lagrange", *Abh. Math. Sem. Univ. Hamburg* 6 (1928), 273-299.

William T. Reid:

[1] "A matrix differential equation of Riccati type", *Amer. J. Math.* 68 (1946), 237-246 and 70 (1948), 460. [MR7, 446 and MR9, 512].

[2] "Oscillation criteria for linear differential systems with complex coefficients", *Pacific J. Math.* 6 (1956), 733-751. [MR18, 898].

[3] "A comparison theorem for self-adjoint differential equations of second order", *Ann. of Math.* (2) 65 (1957), 197-202. [MR19, 1052].

[4] "Principal solutions of non-oscillatory self-adjoint linear differential systems", *Pacific J. Math.* 8 (1958), 147-169. [MR20#4682].

[5] "Riccati matrix differential equations and non-oscillation criteria for associated linear differential systems", *Pacific J. Math.* 13 (1963), 665-685. [MR27#4991].

Ştefan Şandor:

[1] "Sur l'équation différentielle matricielle de type Riccati", *Bull. Math. Soc. Sci. Math. Phys. R.P. Roumaine (N.S.)* 3 (51) (1959), 229-249. [MR23#A1863].

Giovanni Sansone:

[1] "Studi sulle equazioni differenziali lineari omogenee di terzo ordine nel campo reale", *Univ. Nac. Tucumán Revista A* 6 (1948), 195-253. [MR10, 300].

Thomas L. Sherman:

[1] "Conjugate points and simple zeros for ordinary linear differential equations", *Trans. Amer. Math. Soc.* 146 (1969), 397-411. [MR41#572].

C. Sturm:

[1] "Sur les équations différentielles linéaires du second ordre", *J. Math. Pures Appl.* 1 (1836), 106-186.

C.A. Swanson:

[1] *Comparison and oscillation theory of linear differential equations* (Academic Press, New York, London, 1968).

Choy-Tak Taam:

[1] "Non-oscillatory differential equations", *Duke Math. J.* 19 (1952), 493-497. [MR14, 557].

C. de la Vallée Poussin:

[1] "Sur l'équation différentielle linéaire du second ordre", *J. Math. Pures Appl.* (9) 8 (1929), 125-144.

D. Willett:

[1] "On the oscillatory behavior of the solutions of second order linear differential equations", *Ann. Polon. Math.* 21 (1969), 175-194. [MR40#2964].

[2] "Classification of second order linear differential equations with respect to oscillation", *Advances in Math.* 3 (1969), 594-623. [Zbl. Math. 188, 401].

Aurel Wintner:

[1] "A criterion of oscillatory stability", *Quart. Appl. Math.* 7 (1949), 115-117. [MR10, 456].

[2] "On the non-existence of conjugate points", *Amer. J. Math.* 73 (1951), 368-380. [MR13, 37].

James S.W. Wong:

[1] "Oscillation and non-oscillation of solutions of second order linear differential equations with integrable coefficients", *Trans. Amer. Math. Soc.* 144 (1969), 197-215. [MR40#4536].

INDEX

Lecture Notes in Mathematics

Bisher erschienen/Already published

Vol. 38: R. Berger, R. Kiehl, E. Kunz und H.-J. Nastold, Differentialrechnung in der analytischen Geometrie IV, 134 Seiten. 1967. DM 12,–

Vol. 39: Séminaire de Probabilités I. II, 189 pages. 1967. DM 14,–

Vol. 40: J. Tits, Tabellen zu den einfachen Lie Gruppen und ihren Darstellungen. VI, 53 Seiten. 1967. DM 6.80

Vol. 41: A. Grothendieck, Local Cohomology. VI, 106 pages. 1967. DM 10,–

Vol. 42: J. F. Berglund and K. H. Hofmann, Compact Semitopological Semigroups and Weakly Almost Periodic Functions. VI, 160 pages. 1967. DM 12,–

Vol. 43: D. G. Quillen, Homotopical Algebra. VI, 157 pages. 1967. DM 14,–

Vol. 44: K. Urbanik, Lectures on Prediction Theory. IV, 50 pages. 1967. DM 5,80

Vol. 45: A. Wilansky, Topics in Functional Analysis. VI, 102 pages. 1967. DM 9,60

Vol. 46: P. E. Conner, Seminar on Periodic Maps. IV, 116 pages. 1967. DM 10,60

Vol. 47: Reports of the Midwest Category Seminar I. IV, 181 pages. 1967. DM 14,80

Vol. 48: G. de Rham, S. Maumary et M. A. Kervaire, Torsion et Type Simple d'Homotopie. IV, 101 pages. 1967. DM 9,60

Vol. 49: C. Faith, Lectures on Injective Modules and Quotient Rings. XVI, 140 pages. 1967. DM 12,80

Vol. 50: L. Zalcman, Analytic Capacity and Rational Approximation. VI, 155 pages. 1968. DM 13.20

Vol. 51: Séminaire de Probabilités II. IV, 199 pages. 1968. DM 14,–

Vol. 52: D. J. Simms, Lie Groups and Quantum Mechanics. IV, 90 pages. 1968. DM 8,–

Vol. 53: J. Cerf, Sur les difféomorphismes de la sphère de dimension trois ($\Gamma_4 = 0$). XII, 133 pages. 1968. DM 12,–

Vol. 54: G. Shimura, Automorphic Functions and Number Theory. VI, 69 pages. 1968. DM 8,–

Vol. 55: D. Gromoll, W. Klingenberg und W. Meyer, Riemannsche Geometrie im Großen. VI, 287 Seiten. 1968. DM 20,–

Vol. 56: K. Floret und J. Wloka, Einführung in die Theorie der lokalkonvexen Räume. VIII, 194 Seiten. 1968. DM 16,–

Vol. 57: F. Hirzebruch und K. H. Mayer, O (n) Mannigfaltigkeiten, exotische Sphären und Singularitäten. IV, 132 Seiten. 1968. DM 10,80

Vol. 58: Kuramochi Boundaries of Riemann Surfaces. IV, 102 pages. 1968. DM 9,60

Vol. 59: K. Jänich, Differenzierbare G-Mannigfaltigkeiten. VI, 89 Seiten. 1968. DM 8,–

Vol. 60: Seminar on Differential Equations and Dynamical Systems. Edited by G. S. Jones. VI, 106 pages. 1968. DM 9,60

Vol. 61: Reports of the Midwest Category Seminar II. IV, 91 pages. 1968. DM 9,60

Vol. 62: Harish-Chandra, Automorphic Forms on Semisimple Lie Groups X, 138 pages. 1968. DM 14,–

Vol. 63: F. Albrecht, Topics in Control Theory. IV, 65 pages. 1968. DM 6,80

Vol. 64: H. Berens, Interpolationsmethoden zur Behandlung von Approximationsprozessen auf Banachräumen. VI, 90 Seiten. 1968. DM 8,–

Vol. 65: D. Kölzow, Differentiation von Maßen. XII, 102 Seiten. 1968. DM 8,–

Vol. 66: D. Ferus, Totale Absolutkrümmung in Differentialgeometrie und -topologie. VI, 85 Seiten. 1968. DM 8,–

Vol. 67: F. Kamber and P. Tondeur, Flat Manifolds. IV, 53 pages. 1968. DM 5,80

Vol. 68: N. Boboc et P. Mustață, Espaces harmoniques associés aux opérateurs différentiels linéaires du second ordre de type elliptique. VI, 95 pages. 1968. DM 8,60

Vol. 69: Seminar über Potentialtheorie. Herausgegeben von H. Bauer. VI, 180 Seiten. 1968. DM 14,80

Vol. 70: Proceedings of the Summer School in Logic. Edited by M. H. Löb. IV, 331 pages. 1968. DM 20,–

Vol. 71: Séminaire Pierre Lelong (Analyse), Année 1967 – 1968. VI, 190 pages. 1968. DM 14,

Vol. 72: The Syntax and Semantics of Infinitary Languages. Edited by J. Barwise. IV, 268 pages. 1968. DM 18,–

Vol. 73: P. E. Conner, Lectures on the Action of a Finite Group. IV, 123 pages. 1968. DM 10,–

Vol. 74: A. Fröhlich, Formal Groups. IV, 140 pages. 1968. DM 12,–

Vol. 75: G. Lumer, Algèbres de fonctions et espaces de Hardy. VI, 80 pages. 1968. DM 8,–

Vol. 76: R. G. Swan, Algebraic K-Theory. IV, 262 pages. 1968. DM 18,–

Vol. 77: P.-A. Meyer, Processus de Markov: la frontière de Martin. IV, 123 pages. 1968. DM 10,–

Vol. 78: H. Herrlich, Topologische Reflexionen und Coreflexionen. XVI, 166 Seiten. 1968. DM 12,–

Vol. 79: A. Grothendieck, Catégories Cofibrées Additives et Complexe Cotangent Relatif. IV, 167 pages. 1968. DM 12,–

Vol. 80: Seminar on Triples and Categorical Homology Theory. Edited by B. Eckmann. IV, 398 pages. 1969. DM 20,–

Vol. 81: J.-P. Eckmann et M. Guenin, Méthodes Algébriques en Mécanique Statistique. VI, 131 pages. 1969. DM 12,–

Vol. 82: J. Wloka, Grundräume und verallgemeinerte Funktionen. VIII, 131 Seiten. 1969. DM 12,–

Vol. 83: O. Zariski, An Introduction to the Theory of Algebraic Surfaces. IV, 100 pages. 1969. DM 8,–

Vol. 84: H. Lüneburg, Transitive Erweiterungen endlicher Permutationsgruppen. IV, 119 Seiten. 1969. DM 10,–

Vol. 85: P. Cartier et D. Foata, Problèmes combinatoires de commutation et réarrangements. IV, 88 pages. 1969. DM 8,–

Vol. 86: Category Theory, Homology Theory and their Applications I. Edited by P. Hilton. VI, 216 pages. 1969. DM 16,–

Vol. 87: M. Tierney, Categorical Constructions in Stable Homotopy Theory. IV, 65 pages. 1969. DM 6,

Vol. 88: Séminaire de Probabilités III. IV, 229 pages. 1969. DM 18,–

Vol. 89: Probability and Information Theory. Edited by M. Behara, K. Krickeberg and J. Wolfowitz. IV, 256 pages. 1969. DM 18,–

Vol. 90: N. P. Bhatia and O. Hajek, Local Semi-Dynamical Systems. II, 157 pages. 1969. DM 14,–

Vol. 91: N. N. Janenko, Die Zwischenschrittmethode zur Lösung mehrdimensionaler Probleme der mathematischen Physik. VIII, 194 Seiten. 1969. DM 16,80

Vol. 92: Category Theory, Homology Theory and their Applications II. Edited by P. Hilton. V, 308 pages. 1969. DM 20,–

Vol. 93: K. R. Parthasarathy, Multipliers on Locally Compact Groups. III, 54 pages. 1969. DM 5,60

Vol. 94: M. Machover and J. Hirschfeld, Lectures on Non-Standard Analysis. VI, 79 pages. 1969. DM 6,–

Vol. 95: A. S. Troelstra, Principles of Intuitionism. II, 111 pages. 1969. DM 10,–

Vol. 96: H.-B. Brinkmann und D. Puppe, Abelsche und exakte Kategorien, Korrespondenzen. V, 141 Seiten. 1969. DM 10,–

Vol. 97: S. O. Chase and M. E. Sweedler, Hopf Algebras and Galois theory. II, 133 pages. 1969. DM 10,–

Vol. 98: M. Heins, Hardy Classes on Riemann Surfaces. III, 106 pages. 1969. DM 10,–

Vol. 99: Category Theory, Homology Theory and their Applications III. Edited by P. Hilton. IV, 489 pages. 1969. DM 24,–

Vol. 100: M. Artin and B. Mazur, Etale Homotopy. II, 196 Seiten. 1969. DM 12,–

Vol. 101: G. P. Szegö et G. Treccani, Semigruppi di Trasformazioni Multivoche. VI, 177 pages. 1969. DM 14,–

Vol. 102: F. Stummel, Rand- und Eigenwertaufgaben in Sobolewschen Räumen. VIII, 386 Seiten. 1969. DM 20,–

Vol. 103: Lectures in Modern Analysis and Applications I. Edited by C. T. Taam. VII, 162 pages. 1969. DM 12,–

Vol. 104: G. H. Pimbley, Jr., Eigenfunction Branches of Nonlinear Operators and their Bifurcations. II, 128 pages. 1969. DM 10,–

Vol. 105: R. Larsen, The Multiplier Problem. VII, 284 pages. 1969. DM 18,–

Vol. 106: Reports of the Midwest Category Seminar III. Edited by S. Mac Lane. III, 247 pages. 1969. DM 16,–

Vol. 107: A. Peyerimhoff, Lectures on Summability. III, 111 pages. 1969. DM 8,–

Vol. 108: Algebraic K-Theory and its Geometric Applications. Edited by R. M. F. Moss and C. B. Thomas. IV, 86 pages. 1969. DM 6,–

Vol. 109: Conference on the Numerical Solution of Differential Equations. Edited by J. Ll. Morris. VI, 275 pages. 1969. DM 18,–

Vol. 110: The Many Facets of Graph Theory. Edited by G. Chartrand and S. F. Kapoor. VIII, 290 pages. 1969. DM 18,–